The Final Orbit

Apollo and Space Shuttle

Australia's Orroral Valley Space Tracking Station and the End of Ground-based Manned Space Flight Tracking

Philip Clark

PHILIP CLARK

Copyright © 2016 Philip Clark

This edition published by Dreamstone Publishing © 2018

www.dreamstonepublishing.com

National Library of Australia Cataloguing-in-Publication entry

Creator:	Clark, Philip George, author.
Title:	The final orbit: Apollo and Space Shuttle / Philip Clark
	contributors, Jim Thompson [and ten others]
ISBN:	9780987256645 (hardcover case laminate)
Notes:	Includes bibliographical references.
Subjects:	Orroral Valley Tracking Station--History.
	Manned Space Flight Network.
	Apollo 1 (Spacecraft)--History.
	Discovery (Spacecraft)--History.
	Space vehicles--History.
	Manned space flight--History.
Dewey Number:	Dewey Number: 629.454
Published by:	Philip Clark in conjunction with Dreamstone Publishing
	Copyright © 2016 - Philip Clark email: philclark@iprimus.com.au
	First published 2018

Front Cover created by Kim Lambert and Philip Clark, using NASA photos of the Magellanic Cloud and the Space Shuttle Atlantis as a base.

A full list of image credits is at P 105 of this book.

Disclaimer

Orroral in the snow.

Dedication

This book is dedicated to

Erica, my children and

my grandchildren.

Time passes too quickly.

Table of Contents

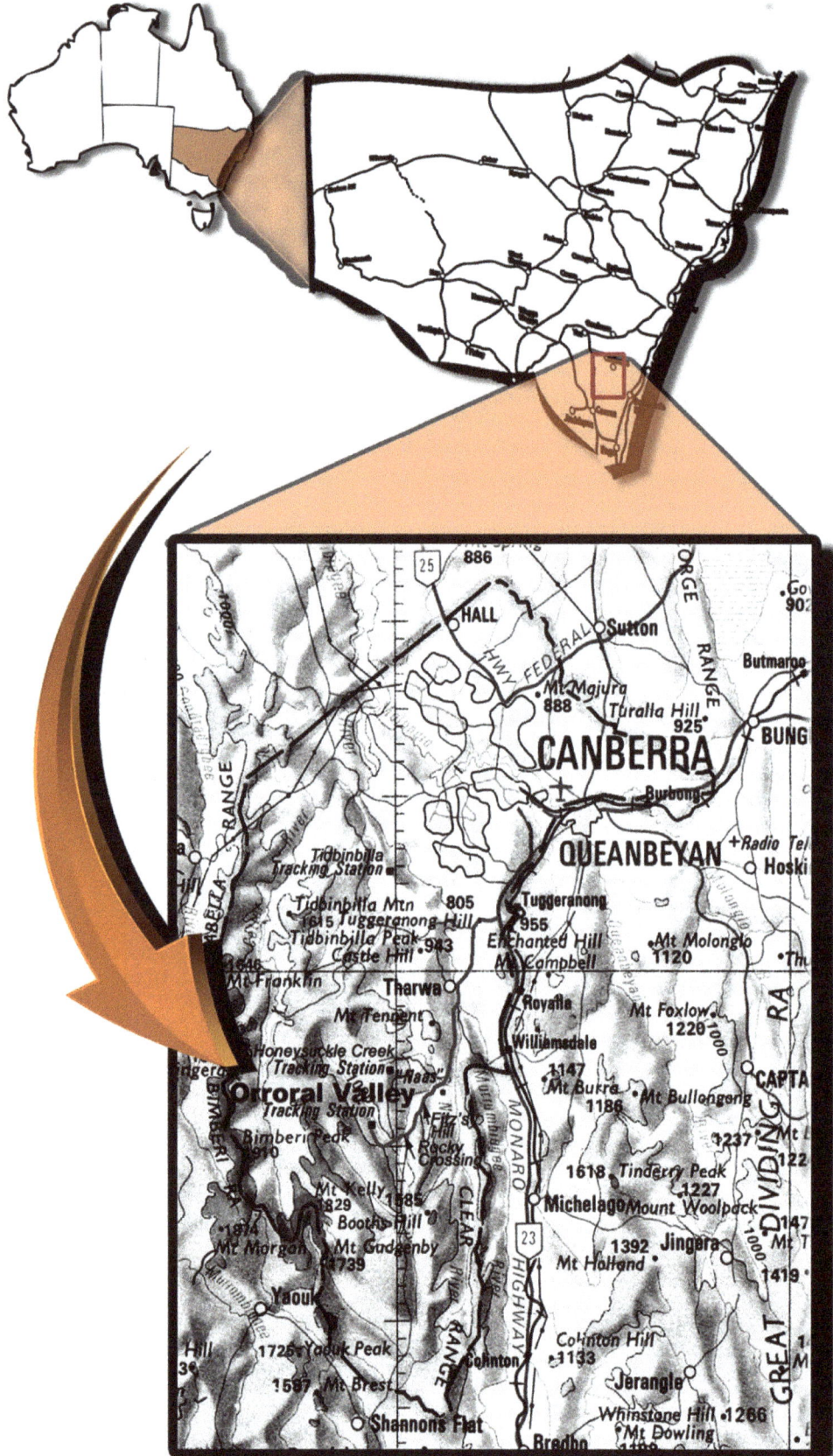

Acknowledgements

I would like to thank the following people for their help and contributions:

Jim Thompson

Hamish Lindsay

Robert (Bob) Quick.

Ross Murray

Peter Uzzell

Richard (Dick) Elliott

Pat Lynch

Lindsay Richmond

Ken Strickland

Thanks to Colin MacKellar for his support and information available from his excellent Honeysuckle Creek Tracking Station Tribute web site (www.honeysucklecreek.net).

I wish to make special mention of the many people who provided me with photographs and/or documents or assisted with identification of former staff. For this reason, attribution has only been given where the originator could be determined with reasonable assurance. Many of the photographs contained in this book have been forwarded from private collections and have been submitted to me on the understanding that they are likely to be published. Much of the pictorial history of the Orroral Valley Space Tracking Station is contained in these collections, and unfortunately comparatively little of this resides in official collections. The author has tried to contact every potential owner of copyright used in this book, known attribution is listed at P105 of this edition, and appropriate attribution will be acknowledged in future editions if so advised.

Excerpts and quotes from interviews with the late Ian Fraser and the late Ian Edgar are used with permission of Hamish Lindsay.

Copies of articles from 'The Canberra Times' are published with permission.

Transcripts of air-ground voice conversations during the Apollo-Soyuz flight are from NASA archives. Transcripts of voice communications of some Space Shuttle flights are from the Author's recordings. All orbital ground track maps are from NASA.

I would like to express my gratitude to Kim Lambert and the team at Dreamstone Publishing for their contribution in putting this publication together.

Philip Clark MSc

Sponsors of This Book

With thanks to our Kickstarter backers, who made this book possible.

Major Sponsor

Erica Clark

Sponsor Level 1

Barbara de la Hunty

Basilio Ormeno

Sponsor Level 2

Noel Crowe

Natasha Chisdes

Colorado Space News

(https://www.coloradospacenews.com)

Sponsor Level 3

Herbert Eder

MichaelAtOz

Michael Uzzell

Lyle Williams

Noel

Sponsor Level 4

Tania Ezra

Ashley

Hugh Blemings

Dik Elliott

Pavel Lihani

Eric Davis

Lindsay Richmond

Dr Andrzej Rososinski

Steve Walsh

Tomasz Ciolek

Mark Bosma

Sponsor Level 5

Paul May

Colin Mackellar

Colin Power

Steve Vincent

Sanjay Ramamurthy

Alan Eisen

Kerrie Dougherty

Steve Howard

Other contributors.

Rob Quick

The Creative Fund

Mark Denbow

Videography services

Kickstarter promotional video filmed and edited by Ryan Schade of ImageFlex Productions.

Preface

Space tracking in Australia dates back to the launch of the very first spacecraft in the late 1950s. Australia had many advantages for tracking spacecraft launched from the United States of America. It even had some initial technical expertise - The Woomera rocket range.

Australia was in an advantageous position. Its geographical location was ideal. Not only was it on the opposite side of the globe from the main United States launch site, its position was such that the early orbits of many satellites launched from the Cape Canaveral site in USA would pass over Australia. This was very important as it enabled the determination of orbital insertion and the actual orbit achieved. Further, Australia was geologically stable, it was politically stable, and it had what was considered a friendly government. This was an important consideration in the 'cold war' years of the 1950s and 1960s.

An agreement between Britain and Australia led to the building of the 'Long Range Weapons Establishment' in South Australia. Len Beadell surveyed the area in 1947, and in 1948 construction crews moved in to build the necessary housing, barracks, and infrastructure. The so-called rocket range was a very large area. The eastern boundary was the railway line that ran between Port Pirie and Alice Springs (the original Ghan railway route) and westward it extended to the South Australia-Western Australia border. The northern boundary was about 180 km south of the border with the Northern Territory. The southern boundary was at the Trans-Australian railway line crossing the Nullarbor Plain above the Great Australian Bight. Occupying nearly one quarter of South Australia (about four times the size of Tasmania) it formed the largest inland test range in the world at that time. The name chosen for the town to support the new range was 'Woomera'. The establishment of this range, later often referred to as Woomera Rocket Range, formed the initial base for space tracking in Australia. In a mutual agreement with Australia prior to the creation of the National Aeronautics and Space Administration (NASA), a Minitrack system had been established at Woomera to track the first United States of America satellite when it was planned to be launched for the International Geophysical Year (IGY) in 1957. But all did not go as planned.

Part of the Minitrack Antenna field.

Left: - Minitrack chart recorders, Jack Manners seated and Ted Peppercorn standing.
Right: - George Riddell at the Minitrack receiver control console with the timing system and power supply units at right. Ted Peppercorn is standing and there is an unidentified person at the table.

The world was astounded when the first artificial satellite was launched by the USSR in late 1957. This was followed by the United States of America in early 1958. Then later there was the excitement as humans were launched into space, first by the USSR then by the United States of America. The surprise launch of Russia's Sputnik in late 1957 meant that even before the Minitrack system at Woomera was used for its designed purpose, it was hurriedly converted to be able to track Sputnik. This was successful and later it was converted back to its original configuration to track the first USA satellite, Explorer 1, which was launched 31 January 1958. Minitrack became the forerunner of the Space Tracking And Data Acquisition Network (STADAN).

In 1960 there was formal exchange of notes between the United States of America and Australia setting out the terms of agreement for space tracking in Australia. This came about because of the need for additional tracking stations and the strong emphasis on manned space flight at that time, for which astronaut selection had begun in 1959. The space tracking agreement between the United States of America and Australia was titled 'Exchange of Notes constituting an Agreement between the Government of Australia and the Government of the United States of America concerning Space Vehicle Tracking and Communication Facilities.'

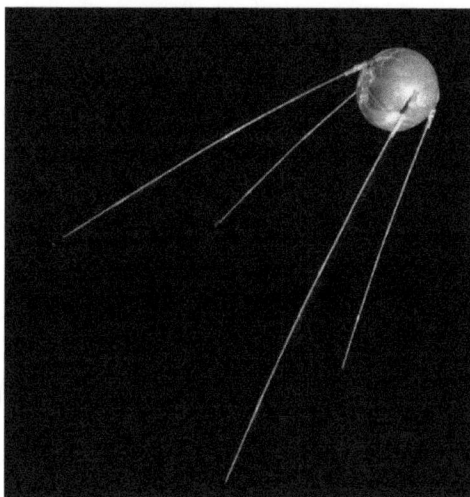

Sputnik.

DSS 41 Woomera

In the early 1960s the United States of America was also beginning to launch deep space probes. For this purpose, they needed a different type of tracking network set up in such a way that the satellites could be tracked for 24 hours a day. To achieve this tracking stations needed to be situated at approximately 120-degree intervals around the Earth. They also needed to be not too distant from the equator.

This was so that they could track a satellite in almost any position of the sky relative to the Earth, taking into account the northern and southern hemispheres. Australia was ideally situated for this purpose being approximately 120 degrees from the United States of America, in the southern hemisphere, and not too far from the equator. The network built for this purpose was the Deep Space Network, DSN.

This book is to provide enlightenment about Australia's vital role in the tracking of manned spacecraft following the closure of the Manned Space Flight Network.

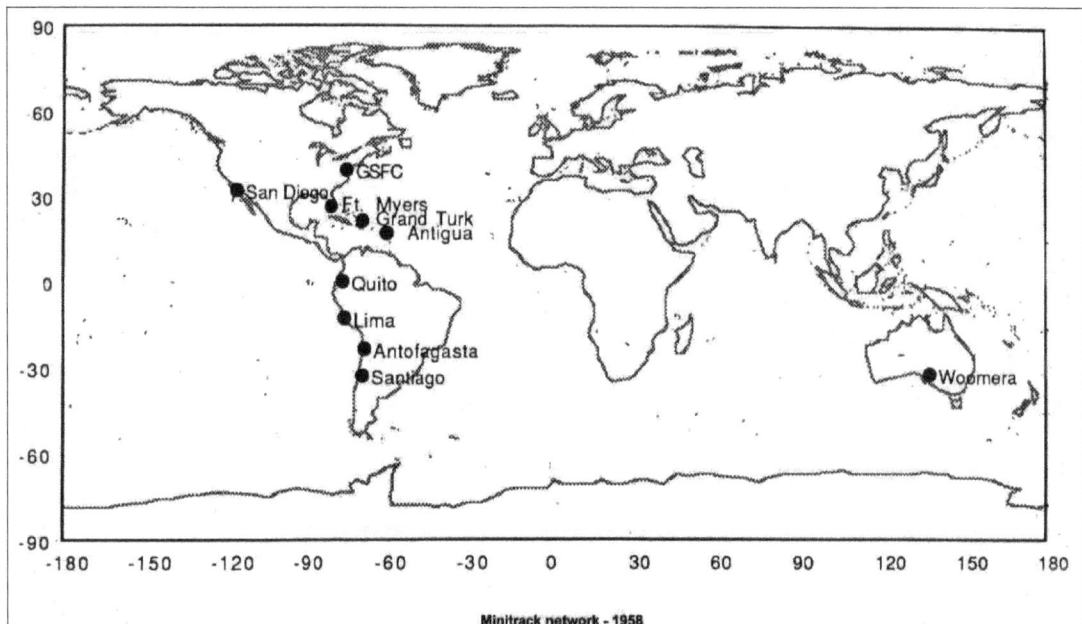

Woomera Tracks U.S. Satellite

ADELAIDE, Sunday. — Woomera's £100,000 mini-track radar this morning tracked the first U.S. satellite on four successive orbits over Australia from 7.15 a.m., 9.21 a.m., 11,28 a,m, and 1.37 p.m.

The first Australian to track the U.S. satellite was a Launceston radio amateur who picked up the satellite's radio signals at 5.27 a.m. as it was heading east-south-east from New Guinea across the South Pacific towards New Zealand.

S.A. "moon watch" will not attempt to observe the satellite through binoclurs or telescopes until the first favourable twilight opportunity occurs an hour or so before dawn in about a week or 10 days.

Dr. Frank Wood, of the Weapons Research Establishment said that the satellite's radio signals were particularly strong when it passed slightly north of Woomera at 7.15 a.m. and equally strong when it passed just south of Woomera on the next orbit at 9.21 a.m.

They were still strong at 11.28 a.m. but were weak at 1.37 p.m. as the satellite neared the equator.

Dr. Wood and Mr. G. M. Bowen, "moon watch" representative of the Wireless Institute of Australia, said that detectors in the satellite's skin transmitted to the earth space information through four small audio oscillators varying in tones according to variations in temperature, etc.

These variations in tone transmitted coded information about the temperature of the satellite's outside skin, the density of cosmic rays the number of small meteorites hitting the satellite and the extent to which small meteorites were wearing the outside of the satellite's nose.

The information was recorded at Woomera and at U.S. earth satellite tracking stations.

The Federal secretary of the wireless institute of Australia, Mr. L. D. Bowie, said in Melbourne that signals from America's "explorer" satellite were heard over most of south-eastern Australia to-day.

Mr. Bowie said that the signals from Explorer's two transmitters were strong. He had heard them for the first time at 5.55 a.m. and again "on a number of occasions" until 2 p.m. when the signals were no longer audible over Victoria.

He expected that signals would again be heard around 6 a.m. to-morrow.

The Explorer transmitted a continuous whistle at a pitch of E above middle C on 108 megacycles. The other transuitter was clearly audible on 108.3 megacycles with a signal which constantly varied in pitch and intensity.

Article from the Adelaide Advertiser 3 February 1958.
From: NLA archives

The perspective in this book is that of the Orroral Valley Tracking Station in the Australian Capital Territory. My aim is to record what we in Australia had and were able to do in support of manned space flight tracking after the Manned Space Flight Network closed.

Because so little was known or recorded about the largest and most diverse space tracking network in the world, STADAN, I felt the need to document this time in history so that people everywhere would know what we did in Australia to support and advance the frontiers of manned space exploration.

A great amount of publicity was given to manned space flight tracking for the Apollo flights to the Moon from 1969 to 1972. At the time it seemed that almost everyone in Australia had heard of the Australian Honeysuckle Creek tracking station and the famous Parkes dish. But very little is known about how manned spacecraft were tracked by ground stations after the final flight to the Moon in 1972. The closure of the Manned Space Flight Network took place in 1972, however it was briefly re-activated from the middle of 1973 until February 1974 to track the trouble-prone Skylab space station. Skylab eventually came to a fiery end in 1979 when it re-entered the atmosphere over south-western Australia.

There is very little public knowledge about how manned spacecraft were tracked by ground stations between the closure of the Manned Space Flight Network (MSFN) and when tracking of Earth-orbiting spacecraft was taken over by the space-based Tracking Data Relay Satellite System (TDRSS) in 1985.

(Author's Note: While the TDRSS did not have its full operational capability in 1985, there was sufficient capacity to allow the closure of most ground stations. After the Orroral Valley tracking station closed in 1985 some Space Shuttle flights were tracked by selected ground stations, including the Tidbinbilla station in the Australian Capital Territory, mainly as a backup to the TDRSS. However, flights that carried classified military payloads and used encrypted data and were not permitted to be tracked from Australia. NASA declared the TDRSS fully operational on 2 July 1989, and after that only a few tracking stations that were needed to support Space Shuttle launches and landings were kept open. The author believes this was the time that tracking of the Space Shuttle from all Australian tracking stations ceased. There were 15 Space Shuttle launches that took place between the closure of Orroral Valley and the TDRSS being declared operational. Of these three were not tracked from Australia because they carried military classified payloads using encrypted data, and one did not achieve orbit due to the Challenger disaster.)

Introduction

The first manned spaceflight tracking stations in Australia were established in 1960 at Muchea near Perth in Western Australia, and at Red Lake, part of the Woomera rocket range in South Australia. These stations supported the one-man spacecraft - Project Mercury, and the later two-man spacecraft - Project Gemini.

Muchea Tracking station.

In 1962 when it was announced that the United States of America would put a man on the moon and return him to Earth, new space tracking networks were needed. At this time space tracking was going ahead in leaps and bounds because the United States of America was a prolific launcher of new spacecraft. They needed more and better tracking facilities and it was decided that they would establish a number of new tracking stations as part of their networks to track these satellites.

New sites in Australia were selected to overcome some of the difficulties that had been found with sites in more remote locations. Such matters were the recruitment, retention and housing of staff, engineering support facilities, transport, communications, and proximity to a university. The Australian Capital Territory (ACT) was chosen as the most suitable place for the new tracking stations.

The first new station to be built in the ACT was for deep space tracking, and this station, established in a valley at Tidbinbilla, near Canberra, took over the Deep Space Network (DSN) tracking from Woomera. The Deep Space Network consisted of three tracking stations spaced approximately equidistantly around the world. Besides Canberra, the other stations were near Madrid in Spain, and Goldstone in the west of the United States of America.

The second Australian station to be built was at Carnarvon in Western Australia for support of the Gemini and Apollo projects. The selection of this site away from the ACT was for a crucial part of the orbit of the Apollo flights to the moon. Construction of this station began early in 1963 just a few months before surveying was started for the construction of the Orroral Valley tracking station.

Carnarvon Tracking Station

The third Australian station was at Orroral Valley in the ACT. This station was part of the Space Tracking And Data Acquisition Network or STADAN.

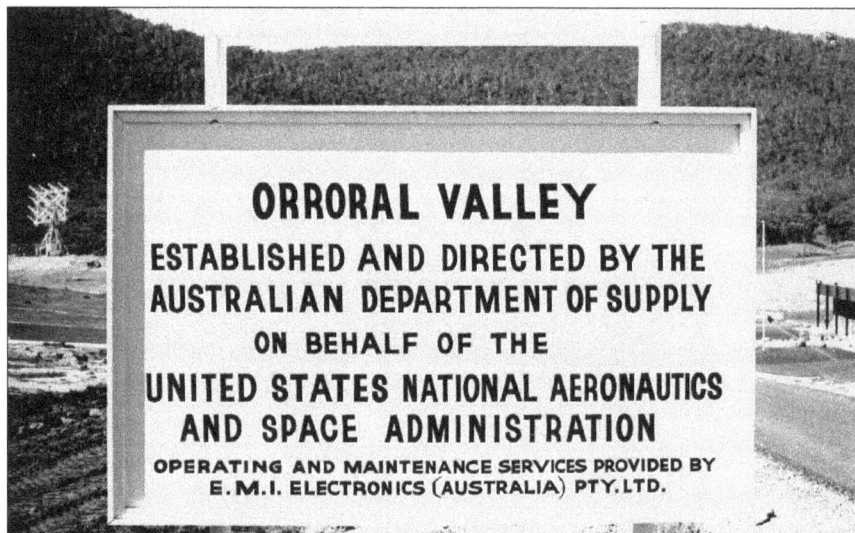

The STADAN network was the largest tracking network in the world. In 1967 the STADAN network consisted of 14 stations, but later in the 1970s, for mainly political reasons, the network had been reduced to 10 stations.

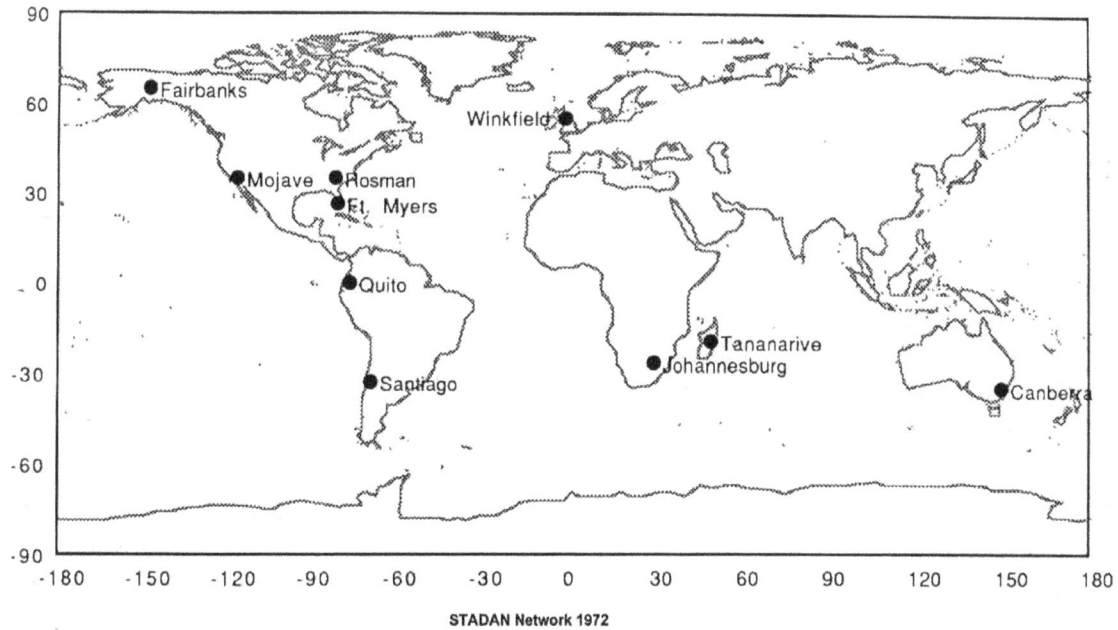

STADAN Network 1972

A fourth Australian station for the Applications Technology Satellite (ATS) project was originally intended also to be in the ACT, but was re-located to Cooby Creek in Queensland because of radio interference considerations.

Project Apollo to the Moon required a radical new communication system. A new concept was devised – the Unified S-Band system, USB, operating in the 1700 MHZ to 2300 MHz microwave band. The USB system combined data, ranging, commands and voice all on a single radio link. However, this brought into play another factor. At the time that the Apollo project was being developed, the maximum data rate that could be sent over normal ground communication circuits to NASA was only about 2,400 to 7,200 bits per second. To support the Apollo data NASA needed to be able to send and receive data at least at 150 kilobits per second. In order to achieve this, NASA used some of the first high speed data Modems available. These Modems operated at speed of 56 kilobits per second, which was quite astounding at the time. The modems were so large that each one needed nearly a whole six-foot-high rack.

The way that NASA engineered the data transmission was quite clever. It was done by splitting the data into smaller packets and then using a 642B computer system to send these packets in sequence over three separate data circuits operating at 56 kilobits per second. Another computer at the receiving end reformatted the packets back into a continuous data stream.

A change in the management and operation of the Australian space tracking stations also took place during this period. Because of the number of new staff required, it was found that the public service procedures were too cumbersome, costly, and slow for the task of recruiting so many new staff. The Australian government agency managing the tracking stations, the Department of Supply (DOS), decided that instead of using public service people to staff the new stations they would adopt a staffing and operating system similar to that used in the USA.

This method was to contract the operation and maintenance of each station out to private companies by tender. The successful contractor then employed the staff and operated the station under DOS supervision. The first contractor at Orroral Valley was a partnership between the Australian arms of EMI Electronics and Hunting Engineering.

People were initially employed under individual contracts, then various ad-hoc industrial awards, until an all-inclusive Space Tracking Industry Award was eventually established. The first employment contract offered to the author is shown in appendix 3 (at P 132). Separate companies tendered individually for the initial operation of the other space tracking stations in Australia.

The Manned Space Flight Network, MSFN, which began with the Mercury and Gemini stations, was upgraded to a dedicated network in 1967. In Australia this led to the building of the Honeysuckle Creek tracking station in the ACT, over a mountain ridge just north of the Orroral Valley station.

The new upgraded MSFN consisted of three main stations with large 26 metre dish antennas. Honeysuckle Creek near Canberra in Australia, Goldstone, in the Mohave Desert of USA, and one near Madrid in Spain. Another station, Carnarvon on the west coast of Australia, with a smaller nine metre dish antenna, played a vital role for Apollo orbit determination and trans-lunar injection. Its location was specifically chosen for these vital functions.

Tracking earth-orbiting spacecraft from the ground was a problem for flight planners. While unmanned scientific satellites could tolerate periods without ground contact, this was not acceptable for manned spaceflight. For example, the STADAN network, even with its some 14 stations, could still only provide about 15% to 20% coverage depending on the orbit height. Spacecraft in higher orbits were able to have better coverage. Once a spacecraft was more than about 40,000 Km from earth, almost continuous coverage could be provided by only 3 stations spaced at approximately 120-degree intervals around the globe. Because NASA placed such a high priority on speaking to the astronauts this was the prime reason for the amateur radio experiment described in chapter 9.

To achieve satisfactory coverage of manned spacecraft while in earth orbit, many additional tracking resources were needed. These came in the form of extra smaller tracking stations, ships, and aircraft.

NASA planned to disband the Manned Space Flight Network at the end of the Apollo moon flights, and this occurred after Apollo 17 in 1972. (Except for the brief period from mid-1973 to February 1974 to track Skylab.) The Honeysuckle Creek tracking station was then transferred to become part of the Deep Space Network.

The tracking of future manned spaceflight after closure of the MSFN was combined with unmanned scientific satellite tracking within the STADAN network. The name of the network was changed to Space Tracking and Data Network, STDN, to reflect its new role.

In Australia the Orroral Valley tracking station, already one of the largest in the world, now had the additional task of tracking manned spacecraft. The first such mission for Orroral was Apollo-Soyuz in 1975, the joint United States - USSR rendezvous mission. However even as Apollo-Soyuz was being launched, construction had already begun for what was to follow - the Space Transportation System or STS – The Space Shuttle, a reusable craft that could land on a runway like a conventional aircraft. This was a capability no previous space vehicle had achieved.

Chapter One - Orroral Beginnings

The site for the Orroral Valley Space tracking Station was selected, with survey and site works started in 1963, and construction commenced in 1964. The station began tracking operations in the latter part of 1965. The official opening took place in February 1966.

Why was such a remote valley chosen for this tracking station?

Unlike Tidbinbilla and Honeysuckle Creek, which used mainly microwave frequencies for their tracking, the frequencies originally used for tracking at Orroral were mostly in the VHF and UHF radio bands. These radio bands were much more susceptible to interference than the microwave frequencies.

Some of the possible sources of interference at these frequencies were two-way radio systems, television transmissions, FM radio stations, air-traffic control, paging systems and even garage door openers. One important factor was stable ground for the antenna bases. So, this valley was chosen because of the distance from the city, the surrounding mountains providing radio shielding, and stable ground for antenna mounting.

This Australian space tracking station at Orroral Valley was part of the largest space-tracking network in the world, STADAN (Space Tracking And Data Acquisition Network). Orroral Valley was also one of the largest tracking stations. In fact, it was believed to be the largest outside of the United States of America. During one period it had a staff of over 200 people, of which nearly two thirds were tracking operations shifts. It had as many as 10 antennas for receiving and communicating to spacecraft. In addition, there were another two antennas dedicated to Japanese weather satellites, a laser tracking system, a Baker-Nunn camera, and for a time, two antennas for monitoring radio emissions from the planet Jupiter. Orroral was capable of tracking and commanding up to seven satellites at once. Orroral Valley Space Tracking Station supported the large number of scientific satellites that helped develop and support the manned space flight program and other missions. In its later years Orroral Valley Space Tracking Station tracked manned spacecraft including Apollo-Soyuz and the Space Shuttle, and also communicated with the scientific packages left on the moon by the Apollo astronauts (ALSEP).

Orroral's Antennas

26 Metre Multi-band receiving Dish Antenna

6 Metre S-band transmit dish antenna

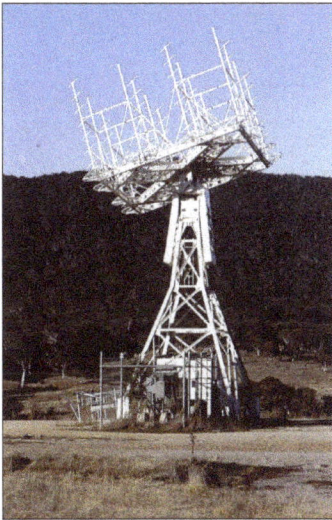

SATAN quad group 136 MHz receive antenna (there were two of these)

136 MHz nine Yagi receive antenna with the 26 M antenna in the background

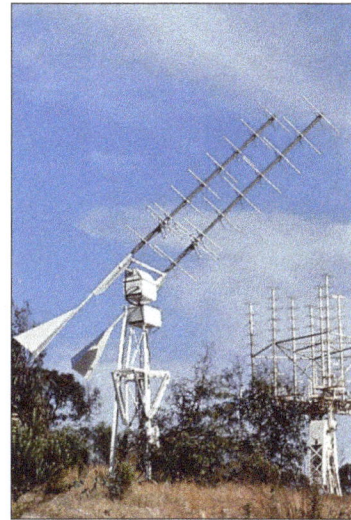

150 Mhz dual Yagi transmit antenna

SATAN Multi Cavity 150 MHz transmit antenna (there were two of these)

9 Metre S-band transmit and receive dish antenna

By 1967 there were three space tracking stations located close together in the Australian Capital Territory. Orroral Valley for the Space Tracking and Data Acquisition Network, Honeysuckle Creek for the Manned Space Flight Network and Tidbinbilla for the Deep Space Network. Both Honeysuckle Creek and Tidbinbilla supported the spacecraft which received almost all of the publicity and media attention. At that time Orroral Valley Space Tracking Station was carrying out some 30 to 70 tracks per day, 24 hours a day and every day of the year. But it received little publicity because it did not support the high-profile space missions that caught the public eye. Orroral did not support deep space probes to far planets and the outer reaches of the solar system and beyond. Not until later in its life did Orroral support manned space flight missions that were more prominent to the public attention.

The Orroral Valley Tracking Station was like a small town with a close-knit population of commuters. There were two main groups of commuters in the Orroral community. These two groups were the Day Workers and the Shiftys or tracking operations staff. The Day Workers did most of the administrative functions and equipment maintenance. The Shiftys were the operations staff who did the actual satellite tracking operations. The Shiftys also did emergency equipment repairs when day maintenance staff were not available. Each group did their commuting in a fleet of cars, mostly with 4 people to a car, although there were some exceptions to this. Day Workers worked from 8:00 am to 4:30 pm each day, Monday to Friday (Senior staff worked slightly different hours.) Shiftys covered 24 hours in 3 shifts each day. There were actually 4 shifts to cover for rest days. On day shift, the Shiftys worked from 8:00 am to 4:00 pm seven days of the week. Considering that the Day Workers and the Shiftys worked mostly apart from one another, there was a quite amazing affinity between all of the people at the station. This was assisted to some extent by an exchange of people between day work and shifts, and also between each of the shifts, as operational and personal circumstances changed from time to time.

Because the station was remote from the city of Canberra it had to be self-supporting in many ways, and so had to have all the essential elements of a small town. If there was a fire, accident, medical, or other emergency, it would have taken the emergency services at least an hour, or more likely longer, simply to reach the station. The emergency service drivers did not know the road as well as the people who worked at the station and who drove it every day.

The Bridge at Naas

Part of the road over Mt. Tennent

The travel times for the emergency services were very much longer than for station vehicles. (Even the police had great difficulty matching the times of station drivers on the unsealed roads!)

There was transport to get people to and from the station, and within the station itself. A fleet of more than 40 motor cars provided this.

Some of the Station car fleet.

Besides standard sedan type vehicles, there were also specialist vehicles such as 4-wheel drives, tractors, trailers, forklifts, 'cherry-pickers', small trucks, and even lawn mowers. Orroral eventually had three 'cherry-pickers, the largest of these had a boom extension of over 25 metres.

A long boom 'Cherry picker'.

A smaller 'Cherry-picker'.

Sometimes flooded rivers added to the driving difficulties.

Part of 'A' Shift being taken to work across the flooded Rocky Crossing

Rocky Crossing in Flood

Mechanics and workshops were needed to maintain these. Storage and pumps were needed for both petrol and diesel fuels for the re-fuelling of all of the vehicles.

The purpose of the station was to track satellites, but it could not do this without the support of all of the sub-sections within the station infrastructure. They were as vital to the operation of the station as its tracking operations area. The antenna structures and the antenna moving parts had to be serviced, painted, and looked after, as well as the cabling and allied electronics, and the electro-hydraulic (servo) systems that moved and guided them. There was even a plumber and carpenter.

There was the electricity supply. The nearest commercial power to the valley was a long distance away through two mountain ranges. Further, because most of the equipment required a 120-volt supply at 60 Hz, the local 240-volt 50 Hz commercial power was not fully compatible. For this reason, the station had to have its own powerhouse complete with staff, reserve fuel supplies and maintenance. Both high and low voltage power was distributed to all parts of the station. Two separate distribution systems were used in the tracking operations area for redundancy.

Three of the diesel generators. In the foreground is a 500 Kw unit, and behind it are two 250 Kw units.

There was a sewerage system. The station had to process its own sewerage in such a way that it was biologically sustainable and did not pollute the surrounding area. It had a three-stage system of settling and aeration ponds.

One of the settling ponds.

Staff on site had to be fed, so kitchen and messing facilities were needed to cover the full 24 hours each day. A team of cooks looked after the provision of meals for all of the station staff on site during the day, and there was a cook on each tracking shift.

The original canteen, built in 1965.

The new canteen, built in 1969.

Hot water had to be provided for air-conditioning and other functions, so there were boiler facilities for this.

Many of the staff were comprehensively trained in first aid as a first line of assistance for any emergencies. Because the station was so far from metropolitan facilities, staff were also trained in firefighting, both for internal building fires and bush fires. The author was a member of one fire team and underwent training with the ACT fire brigade to combat building fires. This meant several sessions using breathing equipment under expert supervision in a smoke chamber. The firefighters were also trained in the fighting of bush fires by the ACT Rural Fire Service. Like a number of other staff members, the author also held advanced qualifications in first aid.

Maintenance of all of the equipment had to be carried out, so a comprehensive store of spare parts was needed. All of this required administration. There was a complete office staff to handle the administrative side of things. External communications were needed. There was a cable connection to the national telephone network that carried external telephone lines and the necessary voice, teletype, and data circuits for connection to Goddard Space Flight Centre. In addition, the station had a separate large internal automatic telephone exchange that could handle up to 1000 connections. The station also had a very extensive public address system that serviced all parts of the station, both internally and externally.

Inside the main operations building there were a number of sub-sections to support maintenance on the various types of specialist equipment used for the actual tracking operations. There was a section to look after the magnetic tape recorders and associated equipment. There was another section to maintain the digital data decommutation and computing equipment. Yet another section oversaw the receivers and the special equipment on the antennas to receive the radio signals, and another looked after the transmitters and their antennas to send commands and instructions to spacecraft. The specialist hydraulics section looked after the drive systems that actually moved the antennas for tracking. There was a section to look after the air-conditioning within the buildings. Another specialist area maintained and calibrated all of the test equipment used for tracking operations, testing, and repair of other systems. There was a separate section that maintained all of the teletype equipment and voice circuit interfaces between the station and NASA in the United States.

A small team of highly qualified engineers supervised all of these functions. At the head of these was the station Supervising Engineer, the highly respected Mr. C.E.D. (Dave) Kemp.

And of course, there was the main operations room(s) that were the heart of the station. It was in here that all the telemetry receivers, tracking receivers, transmitter controls, and their associated equipment were housed and from here the staff could control the antennas and track the satellites as they came within range. (See appendix 4 for information about how satellite signals are sent from the antennas to the receivers.)

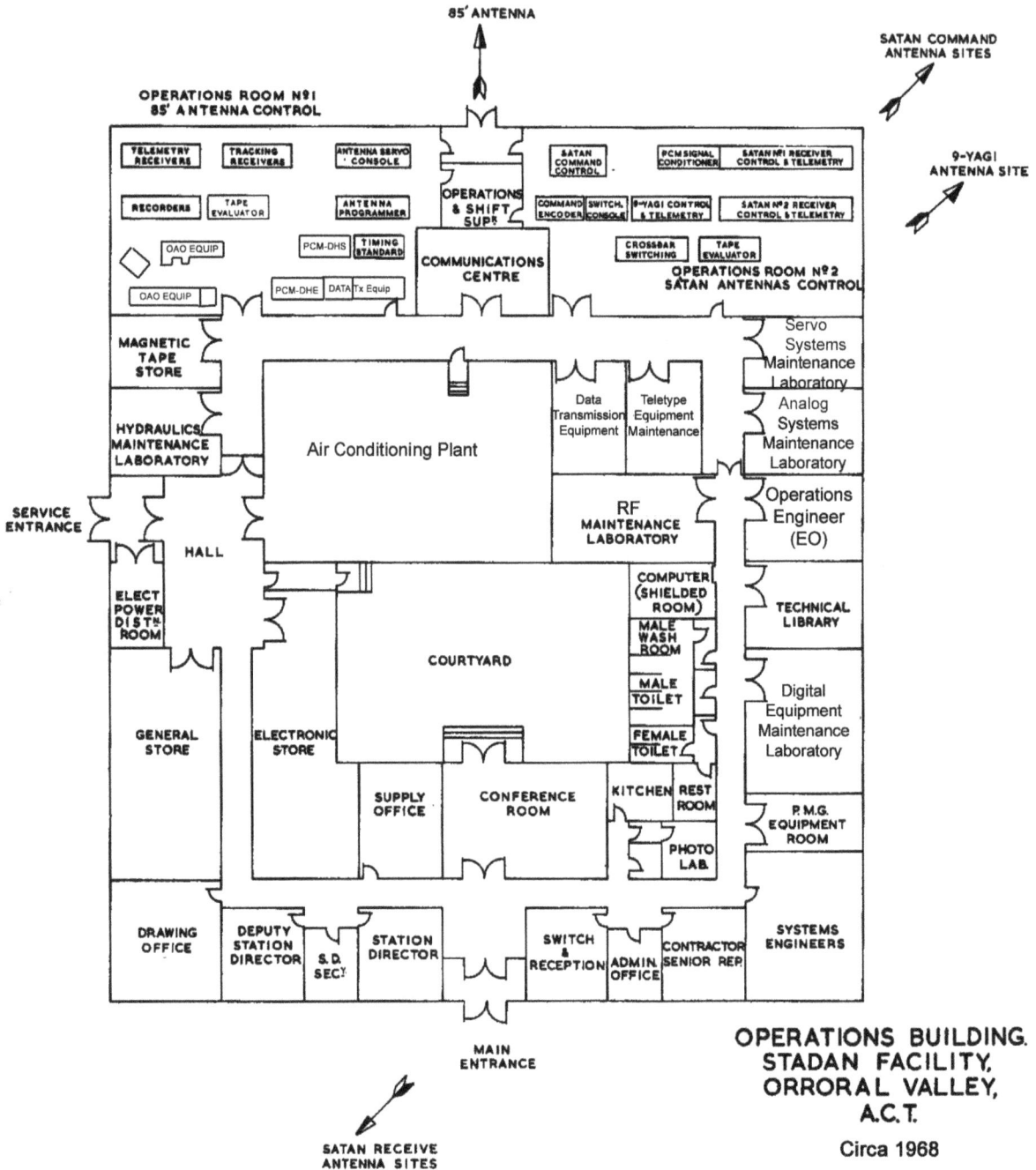

The Orroral operations Building, prior to 1974.

The 26M antenna tracking receivers.

Ted Saar at the AD/ECS computer console

Telemetry receiving system no. 2. In the foreground are two telemetry recorders and an ancillary rack.

The Radiation PCM-DHE and some ancillary racks

A crossbar equipment switching panel
with a configuration patchboard in place above.

The Dynatronics PCM-DHS and the station precision timing system

Telemetry system No. 1 receivers

Telemetry system No. 1 data recorders

In the foreground is a CSC tone command encoder. Gordon Owttrim is seated at the OGO command encoder

The system of operations at Orroral Valley Tracking Station was very different from the stations of the Manned Space Flight and Deep Space networks. The operations at an MSFN station were under the control of the M&O, or Maintenance and Operations supervisor. On some occasions during tracking this position was even staffed by the Station Director. When tracking was being done this position was the prime control and coordination for the tracking operation. It was also the main, and often only, communications point for off-station voice communications to Goddard Space Flight Centre and Houston Mission Control Centre. While the DSN stations had a centralised operations control room, separate from the general equipment area, Orroral did not. Everything at Orroral took place in the two (later one) general operations rooms.

The STADAN stations such as Orroral Valley used a completely different staffing for tracking operations. There was no central operations position or console staffed by an M&O or equivalent as at a MSFN station. There was no separate central operations control room as at the Deep Space stations. The tracking operations at Orroral on each shift were under the overall supervision of the Shift Supervisor, and below him were the operational groups of 'Shiftys' under the less strict supervision of the two Operations Coordinators. An Operations Coordinator was a senior technician who was responsible to ensure that the equipment in a limited area was correctly configured and ready to support a tracking operation. Their task was technical supervisory but was not the same as the M&O at an MSFN station. There were two Operations Coordinators on each shift for tracking operations at Orroral Valley.

At Orroral Valley the Shift Supervisor's task was largely administrative. Only occasionally did the Shift Supervisor take part in operational activities. Orroral had no hard and fast rule that off-station voice contact with a spacecraft control centre had to be done by any particular person or position. If voice contact was necessary with a spacecraft control centre for a pass this would have done by anyone as appropriate for that particular pass, from the Operations Coordinator down. During a pass each tracking position usually made their own announcements direct to the spacecraft control centre they were working with at the time.

Thus, there was no centralised control and communications point at Orroral Valley Space Tracking Station as there was at the Deep Space and MSFN stations. There was not even a position that was a near equivalent. The Engineer Operations (EO), later the Senior Operations Supervisor (SOS), at Orroral was in charge of all the operational tracking shifts. Although this position was in charge of all the operations tracking personnel it was a day worker administrative position. Except for a few special occasions the Senior Operations Supervisor, almost never took part in actual tracking operations.

The equipment at Orroral had to be completely switched around and re-configured after each satellite pass in time for the next one. (See appendix 5 for information about equipment setup and tracking scheduling.) This had to be done on multiple systems some 40 or more times every day often with as little as 10 minutes or less from the end of one pass to the beginning of the next. While the receivers were mostly used with particular antennas, they could be, and often were, switched to different antennas for operational reasons.

Bob Henson relates his experience of setting up to track:

"When you set up for a pass at Orroral Valley you put your project board in. Then you patched the chart recorders, you patched your data recorders, you patched your receivers, and everything else and your brain tells you exactly what's connected to what and it gives you an overall picture of the pass set-up. Over at Tidbinbilla you sit in front of a computer and you go tap tap tap tap like that, a screen comes up. Oh yeah, it's gone red or it's gone green, tap tap tap tap tap, and at the end of the set up to the pass you are none the wiser about how things are configured. I used to like the mechanical side of Orroral Valley where you got to do it yourself and you then understood the system."

Prior to 1974 a telemetry system consisted of an equipment group that included the antenna tracking controls, together with one or more tracking receivers for automatic tracking, the data receivers and data recording equipment, and other ancillary equipment including the 128 x 128 matrix project board patch-panel system. So a single system was be able to track, receive, and record data from a satellite. The command systems were separate, and could be operated independently by one or up to three technicians.

Much of the equipment used for space tracking at the time Orroral was established was based on first generation transistor technology. These transistors were not as efficient in power usage as those developed later and were more sensitive to heat than the ones that followed. Even though they were much more efficient and cooler than the valve (vacuum or electron tube) technology they replaced they still needed considerable cooling. One of the first things that people noticed when entering the operations area was the noise. Not a loud noise as such, but an all-pervading, penetrating, thrumming, humming sound. This came from the many cooling fans in the equipment racks. Most racks had at least one fan and some had many, in either the rack itself or in each of the equipment bays. There were hundreds of fans in total in all of the equipment racks. This completely filled the operations area with a continuous background of sound. In some parts of the operations area the sound was sufficiently loud to make conversation at any sort of distance quite difficult. It may have been this continuous background sound level that contributed to the problems that some people had in coping with shift work over time. The shift workers in the operations area were exposed to the noise continuously.

Information about spacecraft positions was regularly sent from the Goddard Space Flight Centre by teletype. These were called predictions, often abbreviated to 'predicts'. The punched tapes from the teletype machines in the communications centre were then taken and fed into an early model Packard-Bell PB250 computer.

This computer re-calculated and converted the antenna pointing information into a format that could more easily be used by people. This punched paper tape could both be printed out on a teletype machine for an operator to use for manual tracking, or also fed directly into the TE-404 computer to guide the 26-metre antenna under program control. This computer sounded like a clock as each second it 'ticked' to read a frame of data from the paper program tape.

```
05.043200.007834.215.28-46-45
05.043300.007973.217.26-50-44
05.043400.008117.219.24-53-44
05.043500.008266.220.23-56-44
05.043600.008420.222.21-60-43
05.043700.008578.223.19-63-43
05.043800.008740.224.17-66-42
```

This is a printout of some predicted antenna angles in one-minute increments. These predictions are for a spacecraft moving quite slowly relative to the station. Each line reads from left to right as, Day of month (05), Time (Hour Mins Secs), Slant range (Km), Azimuth (Degrees from north), Elevation (Degrees from horizontal), X angle (+ or - degrees from vertical), Y angle (+ or - degrees from vertical).

Of course, all of the preparation work had little point unless a signal was received from the spacecraft. Unfortunately, the antennas do not behave simply as if they are receiving the signal down a long straight tube. The physics of all antennas mean that they have one area where they are most sensitive and receive the strongest signal: in space tracking this was called the main beam or main lobe. This is where the antenna works correctly.

It is only when the satellite signal is being received in this main beam that the antenna can track automatically. Due to the same laws of physics of antennas, the main beam is not just a single tube as it were, it is surrounded by many lesser circular beams referred to as side lobes. (See figure below)

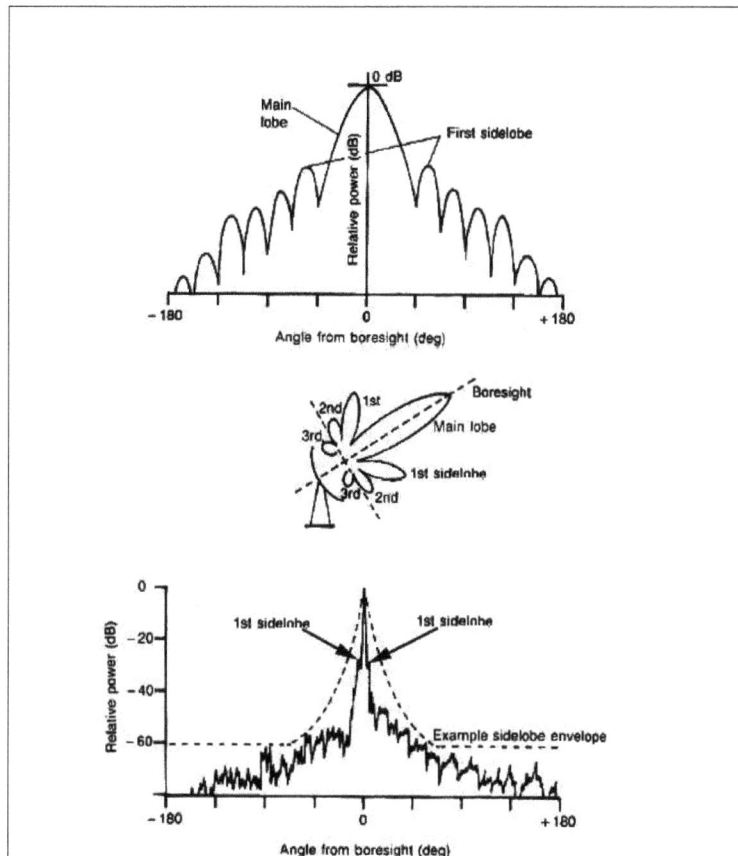

Three different diagrammatic representations of an antenna radiation pattern depicting the main-beam and sidelobes Graphic: ARRL

A signal received in one of the side lobes is not quite as strong as the signal received in the main beam. There are even some side lobes at the very back of the antenna. The further they are removed from the main beam the weaker will be the signal received by them. It is also a characteristic of the antenna physics that the side lobe closest to the main beam has properties such that the tracking error signals generated for satellite tracking are exactly opposite to those required for automatic tracking. Therefore, if the antenna is placed in autotrack mode when the signal is in this side lobe, instead of the antenna trying to track towards the satellite it will drive in the opposite direction and completely lose the signal. This is a total disaster for receiving any data from the satellite.

Unfortunately for the antenna driver the side lobes were very easily mistaken for the main beam unless the driver was skilled and very careful. When the large 26 metre antenna was tracking a satellite using the VHF 136 MHz frequencies, the main beam was approximately three degrees wide and the side lobes correspondingly further away. At the 400 MHz frequency the main beam was approximately one degree wide and the side lobes closer to it. However, in later years when the antenna was using S-Band, the main beam was only about 0.2 of a degree wide and the side lobes very close. This meant there was little or no margin for error when directing the antenna to a spacecraft that used S-Band signals. It was a tribute to the skill of the operators using the antenna at these frequencies that they were able to accurately position the antenna on the satellite to be able to track automatically.

If the predicted angles for the satellite were not sufficiently accurate then it was unlikely that the antenna operator could position the antenna to acquire the satellite in the main beam. In some cases, the antenna operators had local knowledge about a satellite, and if they knew it was early or late on the predictions they made their own corrections. Most times they were surprisingly accurate at this.

Hugh Cocking expressed his feelings about operations:

"I most enjoyed the real-time operations. The satisfaction of waiting for the satellite to come up and make sure everything was all right, and make sure everything went all right, then getting through the pass.

Whether it was a standard one or anything else it was just good to finish satisfactorily. I really enjoyed it. In those early days in the sixties and seventies it was all hands-on stuff, later on when the computers came in and things were more dependent on computers it was not quite so satisfying."

The author recalls one experience not long after he first went to work at the station. Some satellites would pass over the station very quickly and their orbit was such that it would take them virtually straight along the Orroral Valley. One in particular was called Pegasus. This was a satellite designed to measure micrometeorites near the earth.

Seeing a large antenna moving at high speed to track these satellites was an awe-inspiring experience. The author recalled watching the antenna from the control console for some previous passes. Because the 26-metre antenna was limited to a maximum speed of three degrees per second, this spacecraft moved so quickly that the antenna could not quite keep up. The antenna tracking error indicators showed the antenna actually lagged slightly behind the spacecraft as it came to the point of closest approach (PCA) to the station. During one of these passes I went outside to watch and hear the antenna tracking the spacecraft.

As the satellite came over the horizon the antenna hydraulic drive motors could be heard to be increasing speed as the spacecraft approached the station. As the spacecraft reached the point of closest approach and was almost directly above the station, the hydraulic motors on the antenna were screaming like a jet engine to keep the antenna moving fast enough to try and keep up with the spacecraft. The whole antenna structure could be seen to be shaking with the movement. It was a really incredible experience. Then as the spacecraft approached the far horizon the motors gradually slowed down until eventually the antenna was again stationary.

WRESAT undergoing compatibility testing.

In 1967 a historic event involving the Orroral Valley Tracking Station occurred. In late 1966 Australian scientists of the Weapons Research Establishment (WRE) at Woomera were offered the use of a surplus USA Redstone rocket. They decided that Australia would launch a satellite of its own.

In early 1967 design work on Australia's first satellite, WRESAT, began in Adelaide, South Australia. This was a joint venture between the Weapons Research Establishment, and the Physics Department of the University of Adelaide. The project was conducted in association with the United States Department of Defence and NASA. In only 11 months the satellite was constructed in Adelaide, brought to Orroral Tracking Station for compatibility testing, then taken to Woomera in South Australia for launch. The rocket had been modified by Australian technicians who added extra stages to achieve orbit and constructed mating to the WRESAT. WRESAT was launched from Woomera on 29 November 1967.

It was tracked by Orroral Tracking Station, together with other stations around the world, and was completely successful. WRESAT re-entered the Earth's atmosphere over the Atlantic Ocean west of Ireland on 10 January 1968.

This was a very significant event that placed Australia among the leaders of space exploration at that time. Australia became only the THIRD nation in the world, after the USSR and USA, to launch their own satellite from their own territory. (There was another launched by France before WRESAT, but it was launched from Algeria and not from French home soil.) It is disappointing that Australia was not able to maintain its momentum in space exploration into the future.

As for all of the tracking stations in Australia, Orroral's voice and data circuits were connected through the NASCOM switching centre in the Canberra suburb of Deakin. At Deakin were the connections to the long-haul circuits to Voice Control and Data Link at Goddard Space Flight Centre in the USA.

Voice Control handled the connections to all of the centres and stations world-wide, something like a manual telephone exchange, but with computer-controlled push button selection. Voice Control was first set up to handle about 100 lines and was later expanded to 220 circuits. The Voice Control centre could be run by two operators and all connections could be made either talk and listen, or listen only. The Data Link section at Goddard Space Flight Centre controlled the configuration of the data circuits between the stations and the satellite control centres.

Kevyn Westbrook operating the Switching Centre at Deakin in the ACT

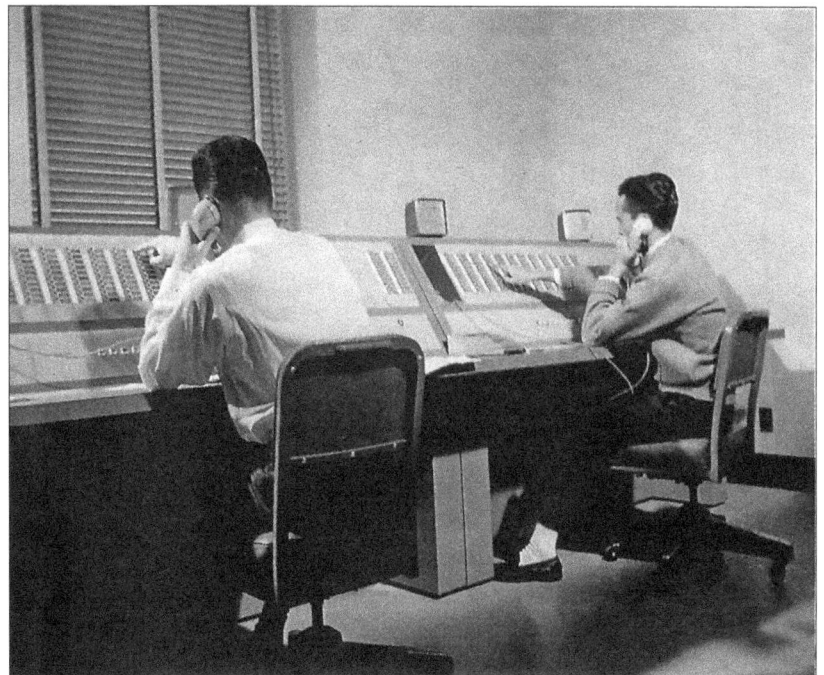

Voice Control at Goddard Space Flight Centre, USA.

Chapter Two - Orroral Prepares for Manned Space Flight Tracking – The New and Upgraded Equipment.

The prime task for Orroral Valley during the Apollo-Soyuz flight was to track the Apollo spacecraft. Orroral station had to be given additional capabilities in order to support manned space flight, as well as a new generation of scientific satellites that were coming on line.

In 1974 the station underwent a major upgrade to meet the requirements for Unified S-Band (USB) support. The operations building was extended to the northeast. An internal wall separating the tracking equipment area from a passageway on the entrance side of the operations rooms was removed, and at the same time, the Shift Supervisor's office, and the communication centre, both of which had been situated in the centre of the operations area, were moved. The communication centre was relocated to a separate room on the entrance side of the operations area and the Shift Supervisor's space was moved to an area near one end of what had previously been Operations Room two. These changes increased the equipment area of the operations area by about 40%. The operations area now became one single very large room packed with equipment. During this upgrade, a 9-metre dish antenna was installed for support of the Apollo-Soyuz and other missions that needed Unified S-Band (USB) capability.

Some former staff members recalled the confusion and chaos while this upgrade work was underway. Many of the floor tiles were lifted so that cables could be added or changed. Staff who walked through the operations area had to be extremely careful not to trip over cables or put their foot through one of the holes where the tiles were lifted. Such an accident could have resulted in a broken leg or worse. Some staff thought that it was one of the hardest times that they had to work through. With so much change and new equipment, staff training was a priority. Unfortunately, in order to accomplish the training efficiently, staff needed to be transferred away from tracking operations for the training. This meant that other staff had to undertake temporary tracking operations duties so that normal scientific satellite tracking could continue. It was a period of considerable turmoil and adjustment.

All of the older GDE telemetry receivers and various types of tracking receivers at the station were replaced with the new multifunction type receivers (MFR) that could carry out both telemetry and tracking functions. Additional receivers were added to the existing systems, giving greater capability, and more receivers were added for the 9-metre antenna system, giving Orroral a total of about 20 receivers. These additions nearly doubled the receiving capacity. All receivers had the capability to be switched to any of the antennas on the station. New transmitters were installed to support the S-Band uplink requirements and other functions, such as ranging.

Rebuilding and extending the operations room, 1974

OPERATIONS ROOM

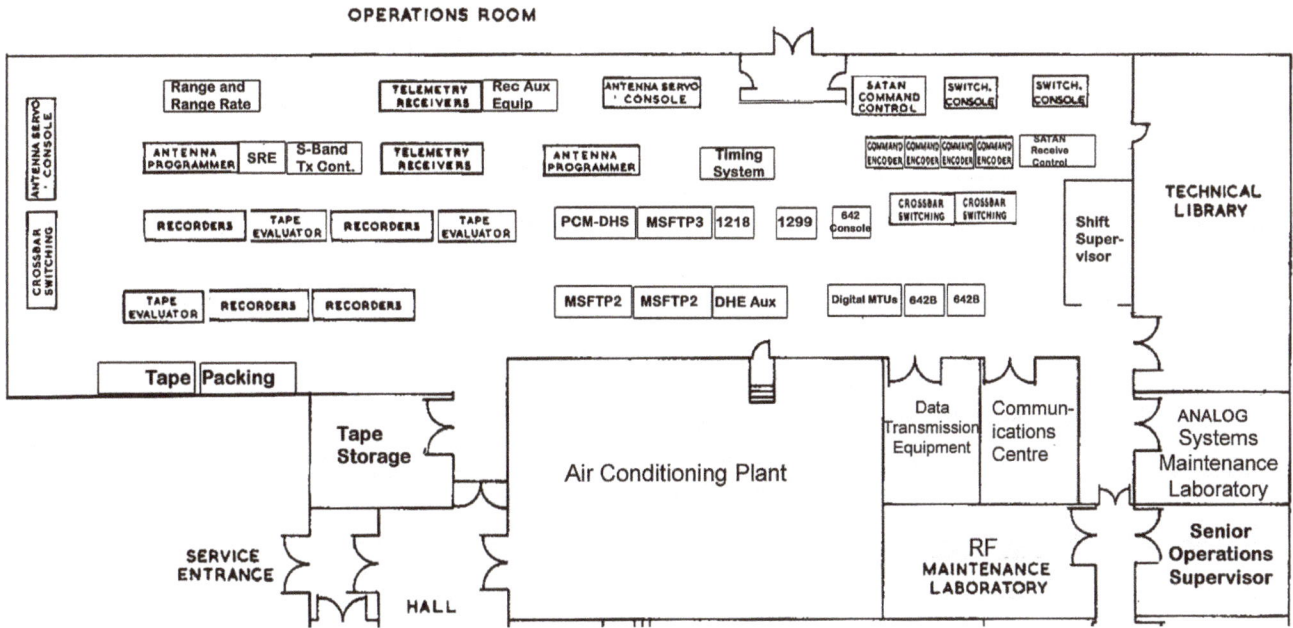

The operations room after 1974 modifications.

During the station upgrade the number of data tape recorders was increased from 10 to 12. The older model Ampex FR-600 tape recorders were retired and replaced with more capable Ampex FR-1100 models and Bell & Howell 8007 types.

Equipment operator Ed Maly loading tape onto a B & H 8007 data recorder.

In the data handling area, the older Radiation PCM-DHE data decommutation equipment was replaced with a Manned Space Flight Telemetry Processor (MSFTP-2) system brought across from Honeysuckle Creek. In addition, two other MSFTP systems were transferred from Honeysuckle Creek, an extra MSFTP–2 and a later model MSFTP–3 type system. This extra equipment doubled the data processing capacity. An S-Band ranging system (SRE) was also installed as part of the 9-metre antenna USB equipment upgrade, to support both ASTP and the later generation of spacecraft.

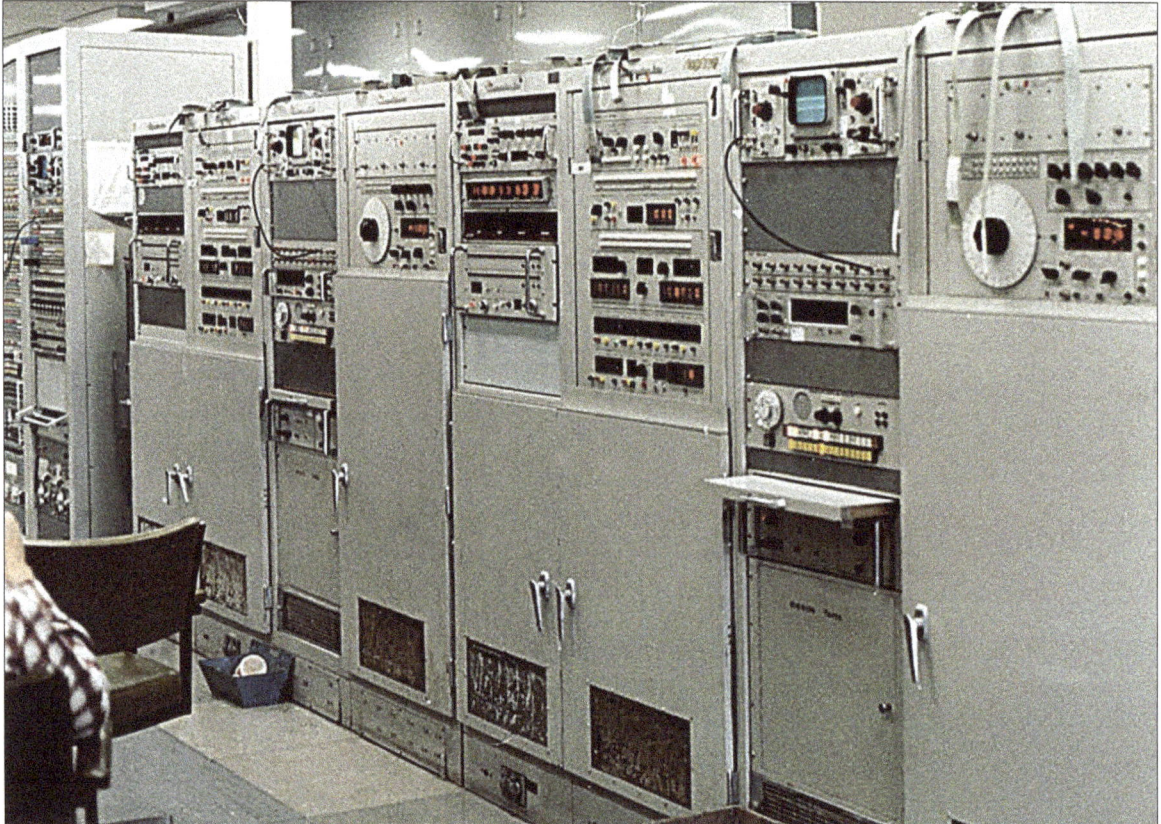

Two Manned Space Flight Telemetry Processors type 2 (MSFTP-2) in the data handling area at Orroral Valley

The older CSC (Consolidated Systems Corporation) tone command encoders dating from the 1950s, and the OGO command encoder of 1960s vintage were removed and replaced with Honeywell computer-based command encoders of the more versatile SCE type. The SCE was based on a Honeywell DDP-316 computer with a physically large (approximately 40 cm / 16-inch diameter) removable Winchester hard disc. Additional SCE systems were also installed so that Orroral could command more spacecraft at once. When this upgrade was complete, Orroral Valley Space Tracking Station could track (receive) up to six satellites and command up to four satellites simultaneously and independently.

All of the new equipment needed to be integrated into the station operational systems. The number of new equipment items added had exceeded the capacity of the existing crossbar switching system. As part of the upgrade several supplementary crossbar switches were installed in the operations area.

With the change to the operations room, the previously separate individual tracking systems were virtually combined. Instead of various components such as recorders, receivers, etc being grouped in each system, they were moved to dedicated areas. (see diagram on P 25)

The receivers were grouped as the 'Receiver Area', command encoders were grouped in what was known as the 'Command Area'. All of the data tape recorders were placed together in what became known as the 'Recorder Area'.

However, for scheduling reasons certain antennas and the associated project board patch panel equipment and crossbar switching panel were locally annotated as a 'telemetry system' and used as such.

Two views of Multi-Function Receivers.

Under this system the 26-metre antenna and the associated project board and crossbar switching was called 'System 1'. Similarly, the 9-metre antenna and associated equipment was called 'System 2', SATAN antenna number one was called 'System 3', SATAN antenna number two was called 'System 4', and the 9-Yagi antenna was called 'System 5'.

The 9-metre antenna. At left is a VHF SATAN receive antenna.

An Operations Supervisor console was installed specifically for ASTP use. This was almost identical to the M&O console of a MSFN station and at Orroral was staffed by the two people trained for the operations supervisor position for ASTP.

At left is a programmable patch board (project board) shown in place above a coaxial patching panel with the cross-bar switch selection panel to the right. Below are storage slots containing project boards for different spacecraft

The S-Band spacecraft ranging system at Orroral.

The screen and keyboard are the operator interface with a computer-based Spacecraft Command Encoder.

The installation of the new equipment and systems was done jointly by a team from Goddard Space Flight Centre working together with station technicians. While all of this work was going on Orroral was still tracking many spacecraft 24 hours a day. It was a tribute to the operations staff that they were able to continue tracking.

The leader of the equipment installation team from Goddard Space Flight Centre, Dave Huff, was a man of impressive dimensions. To the Australians at the station he had a very strong United States accent. He usually had the stub of a cigar poking from his mouth which occasionally would emit smoke with a strong (to some foul) odour. When Dave negotiated the passageways of the building there was no room for anyone else. But he was quite affable, he did get on with most people, and he knew what was required of the job.

Many members of this installation team seemed to project an image of tough guys, however this image was somewhat tarnished when they were confronted by any of the tiger snakes, red bellied black snakes, copperhead snakes, and other snakes that routinely infested the equipment and cable areas of the Orroral building. While most of the Australians were blasé but careful around snakes, the United States installation team people, having heard how venomous Australian snakes were, seemed to be almost terrified by them. (Some of the local staff members who were able to handle snakes were often called upon to remove such intruders from the building.)

Pre-mission planning specified that Orroral would have voice communications capability to the Apollo-Soyuz using their S-Band system, a feature that did not exist previously. The receiver and transmitter systems used by the STDN for tracking the ASTP were quite different from those that had been used at the now decommissioned MSFN stations. The antennas were different, the receivers were different, the command systems were different, and the operational style was different. Equipment to integrate the voice with the uplink command data stream and to decode it from the downlink telemetry data was installed during the 1974 station upgrade. An air-ground voice control console was placed near the data handling area in the operations room.

The station ComTech position. The manned spaceflight voice control console.

This control console had the QUINDAR filters, controls, and switches for the CapCom and astronaut voice transmissions. (The QUINDAR tones are the 'beep beep' sounds heard at the beginning and end of transmission to the astronauts. They were used to control the transmitters for voice communications.) The air-ground console was later upgraded for the Space Shuttle support.

The Apollo-Soyuz flight was a low altitude orbit (only about 225 Km). This orbit meant that the relative speed of the spacecraft with respect to the station was very fast. The antennas had to move at high speed, almost at their maximum permitted rate.

In fact, according to a 'NASA Facts' publication issued just before mission launch, it was considered that the tracking speed would probably be too fast for the larger and heavier 26 metre antennas. Because of this it was planned to support Apollo-Soyuz using only the smaller and more nimble 9 metre antenna systems.

The Operations Supervisor console was installed alongside the 26 M antenna controller TE-404 computer. This console was only used for the ASTP flight and was later removed.

An Operations Supervisor console, similar to this one at Honeysuckle Creek, was used at Orroral only for the ASTP flight.

Before this upgrade the normal data circuits from Orroral operated at a speed of 7.2 kilobits per second. The requirements for support of manned space flight needed higher speed data lines than any Orroral had previously needed. Three lines at a rate of 56 kilobits per second each were required for the whole route between Orroral and both Goddard Space Flight Centre and Houston.

Such data lines had previously been provided to Honeysuckle Creek, however the changeover of support for manned space flights from Honeysuckle to Orroral meant that somehow these high-speed data lines had to be extended to Orroral. Three of these high-speed data circuits were needed with another used as backup.

Two Univac 642-B computers and a Univac 1218 computer were transferred from Honeysuckle Creek tracking Station together with all of their ancillary equipment and digital tape recorders.

The 642B computers (centre) at Honeysuckle Creek before the move to Orroral. The unit in between the two computers with the grey box on top is the extended memory unit. On the far side are the digital magnetic tape units.

The 642B computers at Orroral. Alan Sholtez is seated in front of the extended memory unit. Further along are the digital magnetic tape units.

This computer system included the large 1299 switches that enabled the entire computer system in use to be quickly switched to the backup if required. These computers were used to format the high-speed data stream back to GSFC and the Manned Space Flight Centre at Houston in the USA as previously described.

Another computer at the receiving end reformatted the packets back into a continuous data stream. The 642-B computers also received instructions from both Goddard Space Flight Centre and Houston MCC to send commands to spacecraft.

The communications tower for Australian commercial communications (telephone and other data lines) to Honeysuckle Creek tracking station was positioned on top of the mountain ridge between the two stations, so it was also in view of Orroral Valley Tracking Station. This meant the required data connection was easily provided by installing a microwave link from Orroral to the commercial telecommunications site above Honeysuckle Creek.

Chapter Three – Training for ASTP – APOLLO-SOYUZ Test Project

NASA decided that special training was needed for selected staff from the prime tracking stations which would support Apollo-Soyuz. The stations I recall being selected for training in 1975 were: Bermuda; Santiago, Chile; Orroral, Australia; Quito, Ecuador; Rosman USA; and Tananarive, Malagasy Republic.

(Historical Note: In February 1975 the President of the Malagasy Republic was assassinated and a rebel government took control. Subsequent demands of the rebel government were unacceptable to the United States and they abandoned the Tananarive tracking station before the Apollo-Soyuz flight. The Tananarive station never supported any other spaceflights.)

For the selected team members this special training was conducted during January 1975 in Bermuda, an extremely arduous assignment but someone had to do it. The Orroral staff who trained as specialists for Apollo-Soyuz support were, Ian Fraser, author Philip Clark, Ernie Cook, and Richard (Dick) Simons. Ian Fraser and Dick Simons were trained for what was to be called the Operations Supervisor (Ops Supv) position for Apollo-Soyuz. Ernie Cook and the author were trained as specialists in the data handling and computer areas. The training sessions were conducted on six days of the week (excluding Sunday) and ran from 8:00 o'clock in the mornings to 9:00 o'clock at night. However not all members were required for the whole time during each day.

Special training for the ASTP television system was also undertaken by Pat Lynch at the Madrid station in Spain. Richard (Dick) Simons, the Senior Operations Supervisor, was chosen because he had previous experience with manned space flight operations at Carnarvon tracking station. Although he very rarely took part in normal tracking operations at Orroral, his knowledge of this aspect was thought to be valuable for this particular mission. The other person chosen for the supervisory support position was Ian Fraser. Ian was one of the senior shift supervisors and his technical knowledge of tracking operations was excellent.

I (author) had stayed in New York for a few days before meeting up with Cookie (Ernie Cook). Cookie had a very sharp wit and dry sense of humour. The Pan American 707 flight taking us from New York to Bermuda encountered some quite moderate turbulence over the ocean. Cookie was sipping a drink as a young flight attendant walked past. After a few more bumps he called her over. In a very southern USA accent she politely said to him, "Yes, sir?"

Ernie, with a VERY straight face and worried voice said, "Err, err, Miss, that bloke up front driving, he's not a learrrrner, is he?"

Meanwhile, I was sitting there near the window trying to stop choking with laughter. The young lady then said very seriously in her best southern accent, "Oh no, no, Sir! He is a very experienced pilot."

She had no idea her leg was being verbally pulled, but she seemed to make a point to stay away from Cookie for the rest of the flight. The only unfortunate aspect to this journey was that I had contracted a very bad cold while in New York. I had purchased some 'Vicks' cough medicine in New York to take with me. Bermuda Customs, for whatever reason, confiscated the 'Vicks' when I arrived.

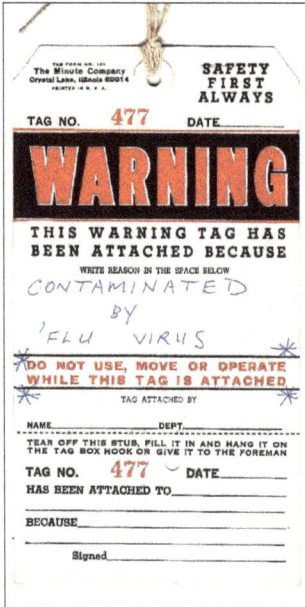

So I was feeling quite ill when I went to bed that night.

On the next morning, a sunny Sunday, I presented feeling and apparently looking very ill. Somehow one of the other people from Orroral on the course, I think it was Cookie, had got hold of a defective equipment tag and he had annotated it with "Contaminated by Flu Virus!". He had said jokingly that I should wear the tag, which I did for the entertainment of all present.

Then I was presented with a tumbler of brown fluid and told that it was medication. It certainly helped! I later found out that the 'medication' was Courvoisier cognac, at the time it seemed this fine beverage was ridiculously cheap in Bermuda.

The Apollo-Soyuz training was conducted at the NASA Bermuda tracking station.

This tracking station, like many other across the world, was operated by mainly United States people of the Bendix Field Engineering company of USA under contract to NASA. However, Australia was different because all tracking stations were operated by all Australian staff.

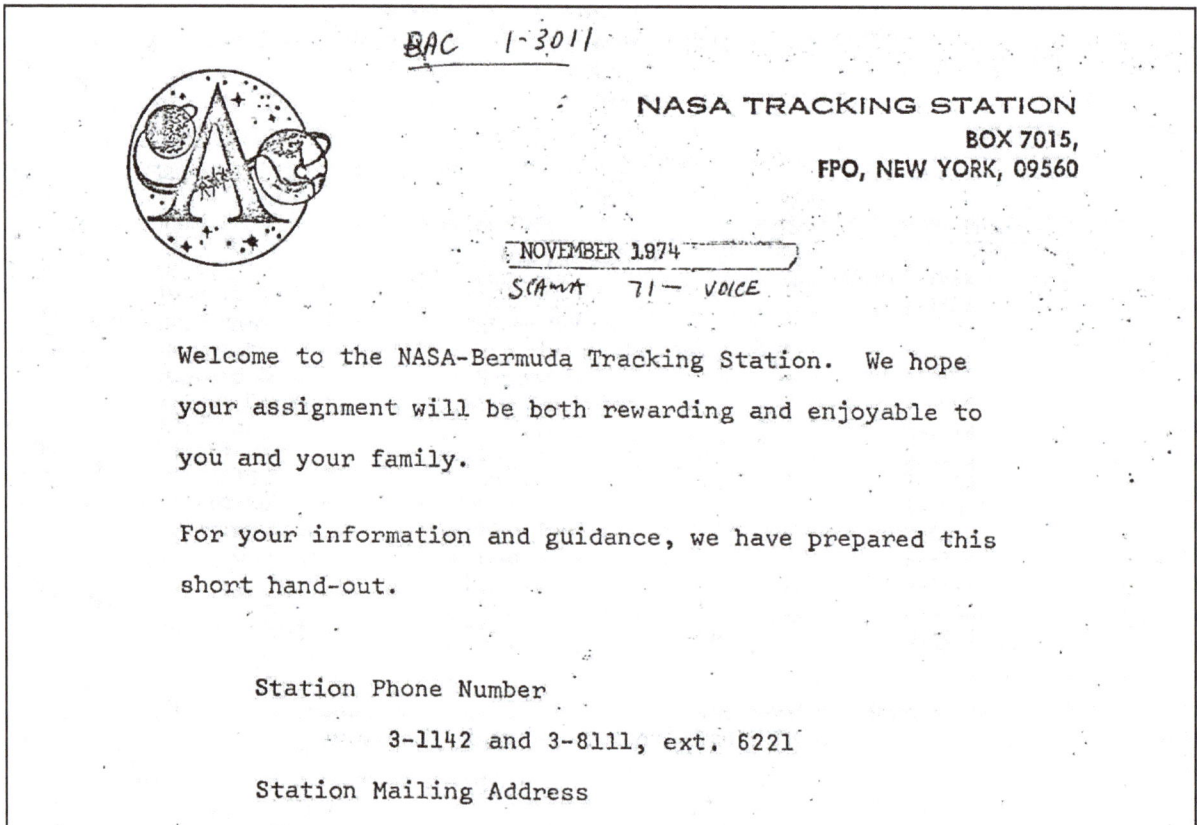

Part of the welcome package given to course attendees on arrival at Bermuda Tracking Station.

The Bermuda tracking station was located on Cooper's Island, off of the tip of St David's Island, at the northern end of Bermuda. Much of St David's Island was taken up by the extensive former US Naval air station, Kindley Field. Access to the tracking station was only through the military air base and subject to strict conditions, such as driving on the wrong (right-hand) side of the road. (Normal driving in Bermuda is on the left of the road.) Part of this airfield was also the Bermuda International Airport until the US forces left Bermuda, when the entire facility became Bermuda's International Airport.

The other interesting aspect about driving to the Bermuda tracking station was crossing the eastern end of the main runway. Traffic lights had been installed so that road traffic could be stopped if an aircraft was taking off or was on approach to land.

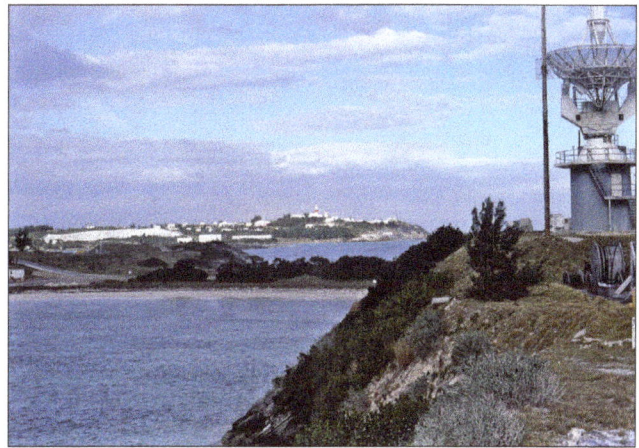

Two views of the FPQ6 long range radar at Bermuda tracking station.

The telemetry building and main antenna of the Bermuda tracking station.

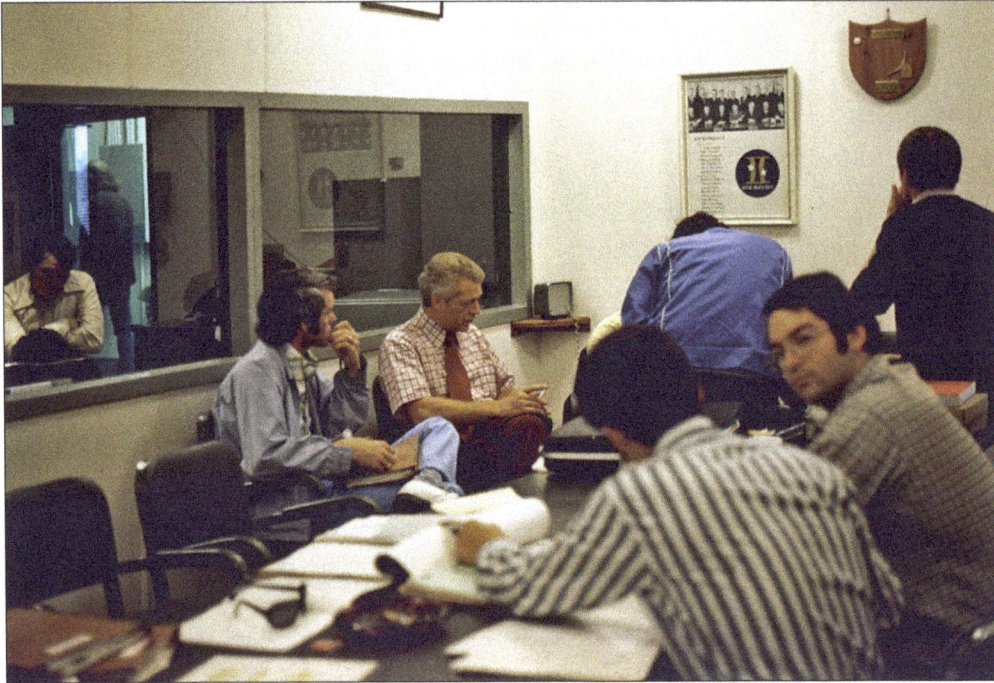

Some of the ASTP course members at Bermuda. Rick Strickland is on the far side of the table at left. William Edmonson is wearing the pink shirt.

The course members were accommodated at the Palmetto Bay Hotel in Flatts village on Harrington sound, about 6 km north of the capital, Hamilton.

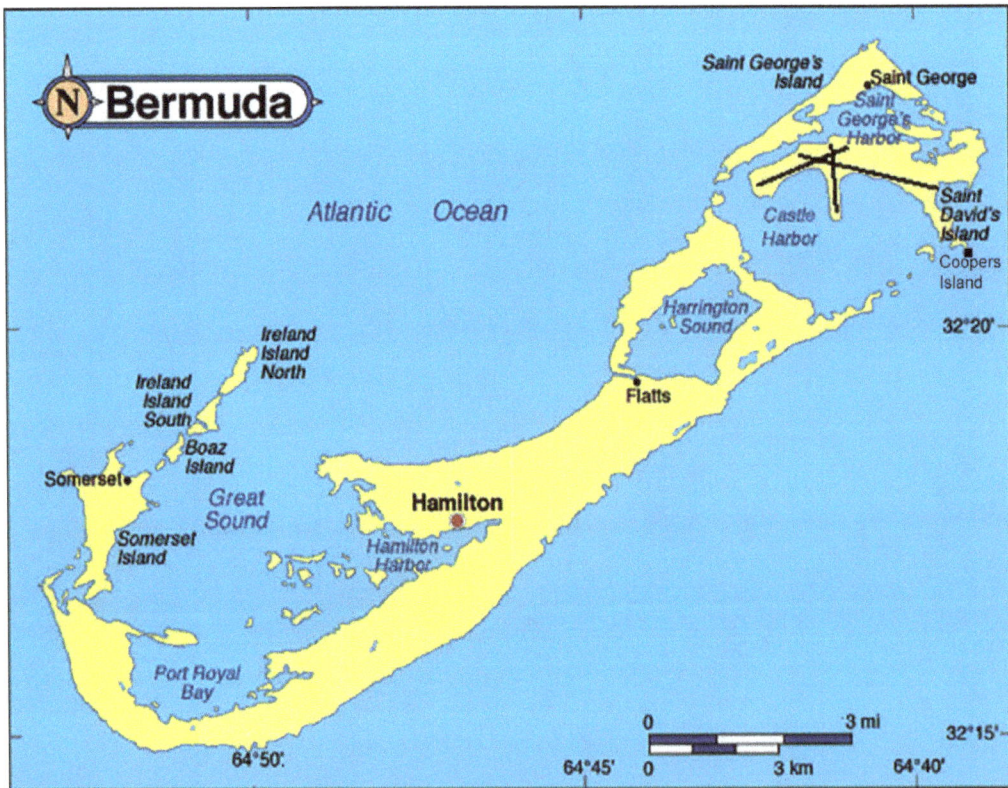

Map showing location of 'Flatts'

The Palmetto Bay Hotel featured an 'Olde English Pub' theme. The serving bar for drinks was only about one metre wide and was named the 'Ha'penny Bar'. The bar top was a glass plate with English halfpenny pieces set underneath. It was supposedly modelled on an old English pub. Forming an arch above the bar and at the sides were about 30 Watney's Brewery half pint glass beer mugs hanging on rings screwed to the woodworked surround. After the ASTP training course was completed there were only about six of these mugs remaining! Nobody seemed to know where they went.

Unfortunately, the rigorous training schedule in sunny Bermuda left little time for the serious business of sightseeing and checking the quality of rum swizzles in the numerous bars and pubs.

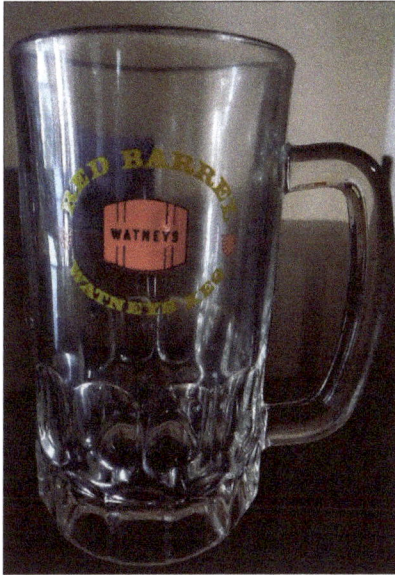

A Watney's beer glass' liberated' from the Palmetto Bay Hotel.

Rick Strickland (left) and another course member relax at the Palmetto Bay Hotel.

The author relates one event that he recalled from the time in Bermuda.

"Aside from the training, which I did not find all that difficult, most of the interesting incidents occurred away from the tracking station. One of these incidents involved a café called 'Clyde's'.

We had been told that some of the best seafood in Bermuda could be had at a café named 'Clyde's' in the town of St George on St George's Island. However, we were told that 'Clyde's' was difficult to find as it was in a back street. A number of us from the course, including all of the people from Orroral, set off to find 'Clyde's' one evening. We had a taxi take us to the town of St George's and the taxi driver told us he knew where 'Clyde's' was. The taxi driver dropped us at the end of a small alleyway and said, "You have to walk down there and you will find the entrance to 'Clyde's' in the wall." The interesting thing was that the entrance to 'Clyde's' was not a full-size door, it was more like a hole in the wall. One did not walk through the door as much as climb through it to get into 'Clyde's', where inside there was the café. The food at 'Clyde's' was absolutely delicious.

Considerable beverages were consumed and everyone was in quite good spirits when we left. The evening that we went out to 'Clyde's' was one of the few times that I had ever seen Dick Simons let his reserve down a little."

With the return of the specialists trained in Bermuda to Orroral Valley, simulated passes were set up to familiarise the normal shift personnel with the ASTP procedures. There was also much new equipment for the operations staff to learn how to set up and operate for tracking. For a while an aircraft was made available for staff training. However, its time was limited and although it gave valuable tracking experience, training had to continue without it.

The training allowed the operations staff to become familiar with the new Multi-Function Receivers (MFRs), the new magnetic data tape recorders, new S-Band transmission and ranging systems, upgraded data handling and processing equipment, and the 642B computer system and its peripherals that had been transferred from Honeysuckle Creek.

Chapter Four - The APOLLO - SOYUZ Flight

At 3:20 p.m. Moscow time on 15 July 1975 an SL-4 launch vehicle lifted Soyuz 19 aloft from the Tyuratam launch site in Kazakhstan USSR with cosmonauts Aleksey Leonov and Valery Kubasov aboard. The launch of Soyuz 19 was the first time that the launch of a Soviet spacecraft was seen around the world on television. Just seven and a half hours later at 3:50 pm Washington time, astronauts Thomas Stafford, Donald (Deke) Slayton and Vance Brand were launched into the Florida sky from pad 39 A at Kennedy Space Centre by a Saturn 1B rocket, for the final flight of Apollo.

Soyuz lifts off from Tyuratam

Apollo lifts of from Kennedy Space Centre.

After the two spacecraft had been launched, the Apollo made a series of manoeuvres that changed its slightly low orbit to bring it into line with the Soyuz. Apollo sighted Soyuz when they were about 400 km apart and then manoeuvred to come closer. It took approximately 2 days of manoeuvring for the two spacecraft to come together.

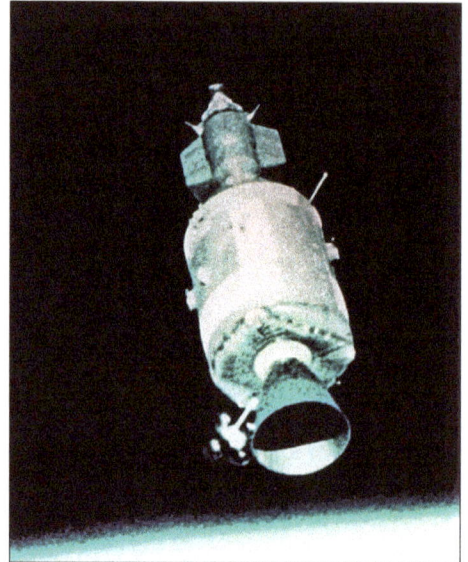

Left: Soyuz photographed from Apollo and Right: Apollo photographed from Soyuz.

Orroral Valley's 9M antenna is tracking

The Orroral tracking station in Australia tracked the Apollo module on all of the orbits that were in range. However, unlike the MSFN stations which usually did only one track at a time, when Orroral was supporting the ASTP mission it still had to track many other spacecraft at the same time.

Orroral was sometimes tracking and commanding up to three other spacecraft at the time it was set up for tracking the Apollo-Soyuz. The workload was sometimes up to as many as 50 passes per day of other satellites.

Ian Fraser put it in context when he said to Hamish Lindsay,

"Apollo-Soyuz was just another spacecraft – except it was manned."

Pat Lynch normally did not work on tracking shifts, but he did have a short sojourn for ASTP as the TV specialist:

"I was one of the Dick Simons' famous Tiger Team. I probably spent most of my time asleep behind the television racks because for the whole of the ASTP mission there were no TV downloads at Orroral. Except I think may be on the very last or second to last pass."

One of the 'firsts' during the Apollo-Soyuz flight was the use of another satellite to track an Earth-orbiting satellite. ATS-6 (ATS - Applications Technology Satellite) was a development satellite for communications. It was moved to a position over Africa for ASTP. The author believes that the use of ATS-6 was in part due to the coup in the Malagasy Republic which caused the abandonment of the Tananarive tracking station. During the Apollo-Soyuz mission, the ATS-6 was used for voice, data communications, and for live television over about 40% of the orbit. Many of the passes at Orroral occurred just after communication with the ATS ended over the Indian Ocean.

The image at left is the ATS-6 being tested before launch.

The Australian tracking stations which had supported the earlier Apollo manned space flights had used a staffing system called 'fluctuating shift'. This system allowed the operations team to adjust their working times to suit the progression or regression of the orbits of the spacecraft, as they changed times day by day. Such a system did not work at Orroral, due to the requirement for 24-hour operations. While taking part in manned spaceflight support, Orroral also continued to work its normal 24-hour 365-day shift roster.

Those staff who had been trained at Bermuda, together with Ian Edgar and Pat Lynch, formed the core Orroral ASTP specialist support team. The specialist staff manned what were considered to be the critical positions. Richard (Dick) Simons and Ian Fraser were at the ASTP operations supervisor console, author Philip Clark was in the data handling area, Ernie Cook was at the 642B computers, Ian Edgar was at the ComTech (air-ground voice) console and Pat Lynch was in the TV area. All other positions were staffed by normal shift personnel.

Former Shift Supervisor and Ops Supervisor for ASTP **Ian Fraser** said to Hamish Lindsay:

"For the Apollo-Soyuz we dedicated a team of operational specialists combined with selected maintenance personnel. Honeysuckle Creek was a 'wing' site with VHF voice support only, providing redundancy in the case of a USB system failure at Orroral Valley." (Honeysuckle Creek staff were on standby to operate this equipment if necessary.)

Paul Hutchinson: Author - I asked Paul what was Honeysuckle's involvement with ASTP.

"At that time, we had the Teltrack antenna and VHF/UHF. We had an uplink and downlink - we had used that for Skylab. I think the whole lot was VHF/UHF, we had a separate building doing the air - ground ComTech".

The Honeysuckle Creek VHF air-ground voice radio system remained on standby for the ASTP mission. Ian Edgar was the Orroral ComTech for ASTP. Ian had been a communications technician (ComTech) at Honeysuckle creek for the earlier Apollo moon missions. He came to Orroral in 1974 for ASTP, and after that mission he remained at Orroral in the digital equipment maintenance section.

Ian Edgar recalled to Hamish Lindsay:

"For Apollo-Soyuz, the station ran the normal four shifts for the scientific satellites, normally about 28 people. There were only four of us operating the equipment for Apollo-Soyuz. We covered 12 to 14 hours with the one shift, we had camp beds set up in the training room, and for that whole period we didn't go home." (Author's note: There were actually six specialists for ASTP: Ian Fraser and Dick Simons as operations and system supervisors, Ernie Cook and the author in data handling and computers, Ian Edgar as ComTech, and Pat Lynch for TV.)

It was a brutal tracking schedule for the specialist team. Because of the period (time) of the orbit, the angle of the orbit, and rotation of the earth, the spacecraft passes came over Orroral tracking station mostly in groups of two, but occasionally three, and because there was usually only a period of about five or six hours between groups, the specialist team had to remain on station for the whole of the mission period.

Pat Lynch's humorous depiction of the author working in the Data Handling area during the ASTP flight.

Such long periods of support were partly due to the pre-pass preparation and System Readiness Test (SRT) which had to be done with Goddard Space Flight Centre and mission control in Houston before each group of passes. Taking some three hours, this SRT was to ensure that the station and its connections to Houston were all 'GO' for the group of passes. This SRT seriously reduced the resting time for the specialist team between groups of passes.

The requirement to carry out the SRT was scheduled on the station daily tracking schedule. This is shown in the later chapter "Tracking Space Shuttle". With less than 90 minutes between passes and the shorter pre-pass preparation of about 30 minutes leading to an H-10 interface (Horizon minus 10 minutes) for the next, there was not much time to spare once the spacecraft orbits were in view of the station.

A wake-up call for Dick Simons as seen by cartoonist Pat Lynch.

It was sometimes quite unnerving for the ASTP team to have a shift change part way through a group of passes. Everything would have been set up, tested, and perhaps one pass done, then the shift would change and there would be a different group of Shiftys to work with. This seemed to be somewhat at odds with the concept of the network confidence testing. It gave the impression that Houston had not come to realise that, unlike the MSFN stations, the STDN stations were tracking other spacecraft all of the time. This meant that some of the confidence testing involved personnel going off duty before the next pass, while those coming on duty had not taken part in the confidence testing.

From the Author's view it seemed that the confidence testing did not appear to have quite the same relevance, as far as the tracking staff were concerned, as it had in the past for the dedicated MSFN stations.

It was even more unsettling when the tracking period covered the 8:00 am local time period on a weekday morning. At that time there was both a normal tracking shift change, and the day maintenance staff came on duty. Suddenly there would be an influx of additional people doing things and starting maintenance or repairs on other equipment in the operations area. The ASTP team had to be especially vigilant to ensure that someone did not decide to start maintenance on equipment which would impact the ASTP support.

While ASTP was in orbit, one of the station internal communication conferences was usually dedicated to the world-wide air-ground voice communications net. This allowed the talking between Houston and the astronauts to be heard continuously in the operations area of Orroral. Unfortunately, when the spacecraft was being tracked at Orroral, those staff actually tracking ASTP could not listen to the astronauts. They were too busy on other voice networks setting up and doing the actual tracking operations. However, staff not involved in other operations had the opportunity to listen to the contact while the Apollo-Soyuz was actually being tracked at Orroral.

Many of the passes scheduled at Orroral were very, very short. Some appeared over the station horizon for as little as 30 seconds. The author recalls that most were only two or three minutes long and the very longest was a little more than five minutes.

The first pass at Orroral was orbit 3 on day 197/01:25 GMT (11:25 am local on 16/7/75). This was a low elevation pass.

The very first pass of a manned spacecraft tracked by Orroral was naturally a tense time. Although it was "Just another spacecraft", it was Orroral's first track of a manned spacecraft, so all of the people involved were extra alert for any problems. Everyone was on their toes to make sure that all was ready and nothing was left to chance.

At Orroral, before the first track of each pass in a group, the station went through a lengthy and complicated setup and test procedure. This was similar to that as shown in a later chapter. Once the three-hour SRT period was completed, the station went into the H-45 (Horizon minus 45 minutes) pre-pass period. This was the final interface with Goddard Space Flight Centre and Manned Space Flight Control Centre (MCC) at Houston before AOS (Acquisition Of Signal). During this time all of the data connections to Goddard Space Flight Centre and Houston were tested and checked using simulated data.

The station ComTech spoke with Houston ComTech on the air-ground voice circuits while they were tested and made ready. Ten sequences of QUINDAR tones were sent and the station ComTech had to verify that all were successful. If any failed then adjustments or repairs were made until the ten attempts were successful. The author's memory is that QUINDAR detectors were quite sensitive, sometimes a little temperamental, and needed some delicate adjusting to get things correct. The pre-pass and SRT procedure had brought all of the station equipment and systems to proper configuration, ready to track the spacecraft. At the conclusion of the pre-pass readiness, and the H-45 countdown, everything was ready for AOS of the spacecraft: The antenna was positioned to the intercept point on the horizon, which was predicted for the appearance of the spacecraft; the station uplink transmitter had been switched on; the tape recorders were running; and the receiver technicians were listening for the first sign of signal from the spacecraft.

(Explanatory Note: The spacecraft would not switch on its own transmitter until it received a signal from the ground. Because the tracking station transmitter was very much more powerful that the one on the spacecraft, it could actually reach a little way over the horizon. So it was likely that the spacecraft transmitter would be switched on just before it came over the station horizon.)

As soon as the receiver technicians heard a signal from the spacecraft, they locked the receivers to the signal and the tracking operation began. Typically, this would be announced to Manned Spacecraft Control as "Houston, Orroral has S-Band AOS."

But when it came time at Orroral for the first actual track of ASTP, something did go wrong!

As with all tracking, the station was sent pointing angle predictions for the orbit of the Apollo, by teletype. These gave the predicted angles for pointing the antenna. However, these predictions were always a little behind time and therefore often had some minor errors. To allow for this the tracking station antenna had a device called an acquisition aid.

This was a small antenna with a wider beam width that was attached to the front of the main antenna. It was supposed to enable the antenna operator to properly point the main antenna at the spacecraft. For some reason, which was never established, the acquisition aid at Orroral did not seem to work properly for ASTP. The first pass at Orroral was low elevation and occurred just after a thruster burn to correct the orbit. The pass was only a little over two minutes duration.

Lindsay Richmond:

"Do you remember the guy they brought over from Western Australia? (Carnarvon) He was the 'anointed' one to drive the antenna, but on the first pass we got it in a side-lobe. We did not get in main beam the whole pass. The next time it came around, we used our own antenna driver and we had no problems. The 'anointed' one never got to drive the antenna again."

Tony Bourne remembered the first pass of Apollo-Soyuz:

"They had it in a side-lobe. They didn't know why they had it in a side-lobe on the antenna. They had Dick Simons and some other ops fellows in the room where the air-conditioning fellow had a big long bench. They had all of these chart recordings stretched out trying to figure out what had happened."

Although the pass was tracked in a side-lobe, all communications were fully supported. As Tony Bourne said, there was an extensive analysis, post pass, to try to determine what the problem was. It was thought that, probably, the combination of the thruster burn and slight angle prediction errors were the cause of the problem. The short duration of the pass did not allow any time to make corrections during the track.

(An antenna has a number of areas where it can receive a signal. The best of these is the main beam or main lobe. Surrounding the main beam are a number of lesser lobes. Although a signal can be received and transmitted through these, the further they are separated from the main lobe or beam the less efficient they become.)

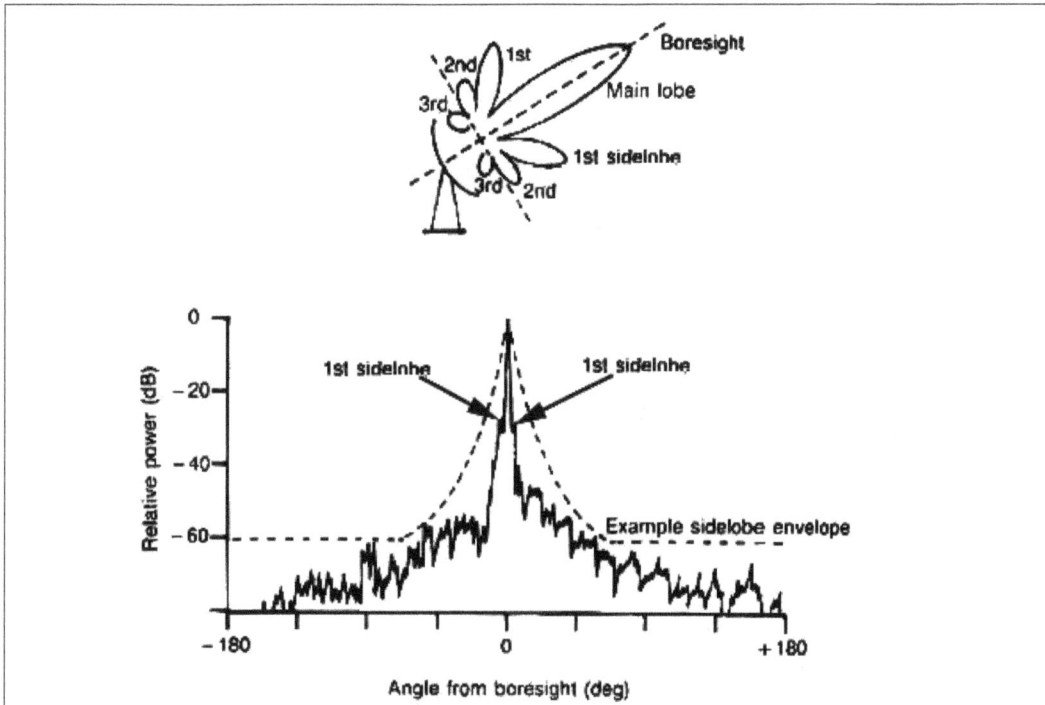

Diagrammatic representations of antenna side-lobes.

As the announcer on the Houston Public Announcement network said at the time:

"This is Apollo Control. The displays here in the Control Centre showed the phasing burn number 1 as it was underway aboard Apollo. However, the loss of signal at Orroral Valley coincided fairly closely with the cut-off of the burn."

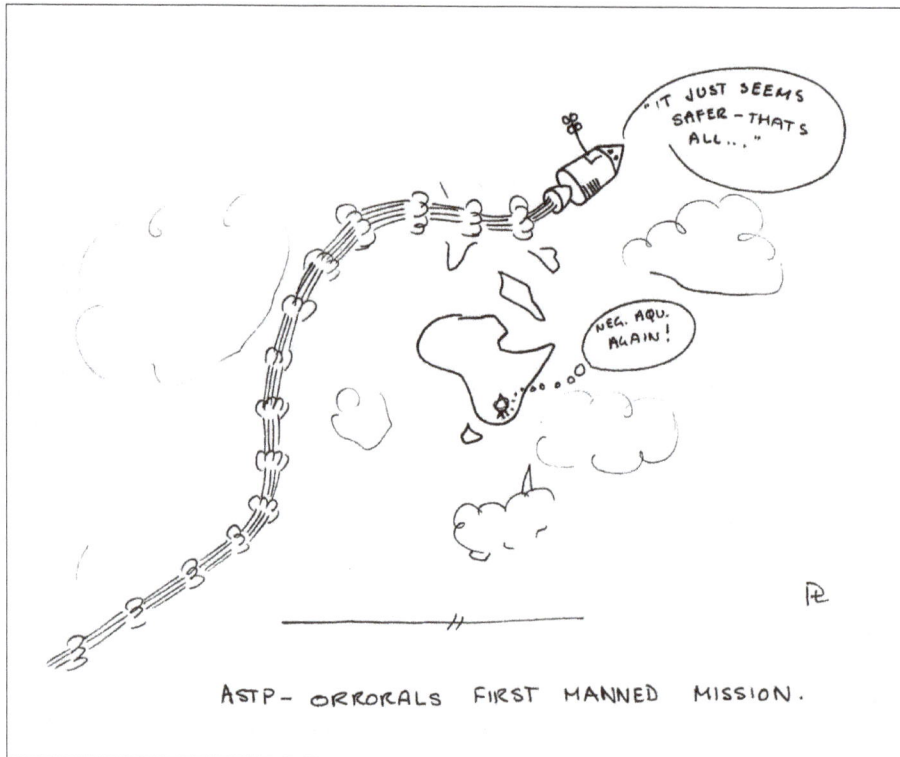

Pat Lynch's cartoon jokingly shows Apollo trying to avoid Orroral for their first manned track.

ASTP mission control, Houston

After a gap of about 5 hours the next pass for Orroral occurred at about 1:45 am in the morning of 17/7/75 and was believed to be orbit 14, a low elevation pass.

Times are GMT.

Day 197 16/7/75

15 46 38	HOUSTON	Apollo, Houston. Short pass at Orroral Valley for about 3 minutes. How do you read?
	APOLLO	Loud and clear, Crip (sic).
	HOUSTON	Roger, Vance. How y'all doing?
	APOLLO	I estimate we are about 20 minutes behind the time line. Deke is just - getting ready to go into the DM.
	HOUSTON	Okay. Super.
	APOLLO	We Just opened hatch 2.
	HOUSTON	Okay. - Thank you.
	HOUSTON	Apollo, Houston. We are about 1 minute from LOS. We're going to have a short Santiago pass at 27:53; then we'll see you on the satellite. The high-gain angles in the checklist are good.
	APOLLO	Okay. Very good, Dick.
15 48 48	HOUSTON	Okay, Vance. See you later.

One orbit later, at 3:17 am on 17/7/75, Apollo passed almost directly overhead the station and this was one of the longest passes supported during the flight.

Day 197 16/7/75

17 16 46	HOUSTON	Apollo, HOUSTON through Orroral Valley for 6 minutes.
17 17 08	HOUSTON	Apollo, HOUSTON through Orroral Valley. We dropped out there for a second. I'm back up and standing by.
	APOLLO	Okay, Dick.
17 21 00	HOUSTON	Apollo, Houston. We're 1 minute from LOS. We'll give you a call at Quito at 29:27. See you there.

Speaking with Hamish Lindsay **Ian Fraser** commented:

"We had trouble finding the spacecraft while they were together but once they separated we had no trouble finding Apollo straight away."

Air-Ground conversation during the following Orroral passes gives the reader some indication of the times the Orroral team was working. With the three-hour SRT period before each group of passes, there was little time for the team to rest. A gap of about six hours saw Apollo back over Orroral Valley at about 9:30 am on 17/7/75 local time.

Day 197 16/7/75

GMT

23 29 41	HOUSTON	Apollo, HOUSTON - through Orroral Valley for 2 minutes.
	APOLLO	Roger, Bo. One item... mention about the VTR coolant activation. Do we still want that, as far as putting the hoses on there?
	HOUSTON	Roger, Apollo. We would like that VTR coolant activation.
	HOUSTON	Apollo, Houston. There is 1 minute until LOS. Next AOS will be Hawaii at 35:26.
	APOLLO	Roger.
	APOLLO	Houston, Apollo.
	HOUSTON	Apollo, Houston. Go ahead.
	APOLLO	Okay, Bo. One question. You know these little ...dosimeters that we had - on our underwear - what do you want done with those, because we also have the big personal dosimeters with counters on them. Do you want these brought back?
23 32 46	HOUSTON	I'll check.

Then again at 11:00 am local time on 17/7/75

Day 198 16/7/75

GMT

01 01 44	HOUSTON	Apollo, HOUSTON - through Orroral for a little over a minute.
	APOLLO	Okay.
	HOUSTON	Apollo, Houston. There are 30 seconds until LOS. We'll say good night to you, and the wake-up time will be as scheduled on the Flight Plan. That's Hawaii at 44:50.
	APOLLO	All righty.
	HOUSTON	Good night.
	APOLLO	Good night.
	APOLLO	See you later.
	HOUSTON	See you later.
01 03 05	APOLLO	Canister's changed.

Richard (Dick) Elliott recalled his experience during ASTP:

"I think I was involved pretty much in just the normal operations while that was on. I can remember some of the passes happening, but at the time there was a sort of Tiger Team, in the same way that I was on the shuttle Tiger Team later. There were other people who were sort of jumping to do it. I was aware of the fact that it was going on, and also we had all the normal passes to be done at the same time on other antennas and other systems."

In conversation with **Hamish Lindsay, Ian Edgar and Ian Fraser** recalled the final pass of the Apollo spacecraft at Orroral on the morning of 25 July 1975:

Edgar: "During that final orbit, at the re-entry stage for the CSM, the main worry we had was what the predicted velocity was going to be..."

Fraser: "...whether the antenna was going to keep up with the speed of the spacecraft. It was so close it was like you could throw stones at it..."

Edgar continues, "...you could almost go outside and watch it - you could actually see it - and it was daylight. Houston told us, 'We don't think you will be able to keep up with it, we're not scheduling data off this re-entry - but should you get it, we'll take it.' To our surprise we did maintain it..."

Fraser: "...the antenna was flat out..."

Edgar: "...it must have been to design limits and didn't break... right through to the horizon. We heard them talking but it started to break up..."

Fraser: "...that's right, the signal started to break up as they came down, but the telemetry stayed okay. We still had the voice on the net [voice line from the USA via ARIA] after we lost them over the horizon, as they were coming in over the tracking ships and the parachutes were deployed, and suddenly everything went quiet. I think the last thing I remember hearing on the net was coughing before somebody pulled the plug."

Lindsay Richmond:

"It was going at something like four degrees a second if I remember. The antenna was having trouble. It was really hiking, very low. It was just on daylight. We wondered if we would even see it (track it) on that pass and get on to it, but we did. After the pass Ian Fraser (Operations Supervisor) produced this bottle of scotch and we all sat around the operations supervisor console drinking scotch, out of plastic cups from the coffee machine, at seven o'clock in the morning. Fraser, in his inimitable style said, "We're having a drink to celebrate!" (Officially drinking alcohol on station was not allowed.) The look on Dick Simon's face would have melted a polar ice-cap but Fraser said "They're my boys and we are going to have a drink!" It was a successful mission as far as we were concerned".

The last pass over Orroral was the re-entry orbit and took place at about 6:53 am on 25/7/75 local time. Apollo entered the atmosphere shortly after LOS at Orroral and the re-entry plasma blacked out communications until the signal was re-acquired by an aircraft over the Pacific Ocean.

Day 205 24/7/75

GMT

20:53:05	HOUSTON	Apollo, Houston. We're AOS through Orroral, on VHF.
	HOUSTON	Apollo, Houston. We're AOS Orroral; have you for a couple of minutes.
	HOUSTON	And, Apollo, be advised that we have nega - no pad updates for you.
	APOLLO	How are you reading us, Crip?
	HOUSTON	Loud and clear, how me?
	APOLLO	Okay. You guys got data down there now?
	HOUSTON	That's affirm. We've got a few minutes coming across Orroral here.
	APOLLO	Okay. How about telling them to take a look at our pyro batteries. I'm showing an amp oscillation. About a minus 2 to plus 2, I don't think it's anything, but - I never noticed it before so they might want to look at them.
	HOUSTON	We'll take a look,
	HOUSTON	Tom - correction, Deke, would you clarify that you're talking about your pyro bus amps?
	APOLLO	That's affirm - pyro. And they should nominally be reading zero - -
	HOUSTON	Rog.
	APOLLO	- but they're just oscillating there, and I'm assuming that's the instrumentation thing, but-
	HOUSTON	That - that's affirmative. They're all safe; there's no problem with them at all.
	APOLLO	Okay.
	HOUSTON	Okay, we're about to go over the hill here at Orroral, and we should have you after blackout through ARIA.
	APOLLO	Roger.

Splashdown of this flight signalled the end of the Apollo era. The next time USA astronauts went into space it would be on the Space Shuttle.

INTERNAL MEMORANDUM   STADAN ORRORAL

TO	S/S, A B C D	YOUR FILE NO.	
FROM	SE(O)	OUR FILE NO.	
COPIES	Messrs Fraser, Cook, Clark, Edgar, Lynch, Green. SD, CE, SE(DA), SE(DH).	DATE	25 July, 1975.

SUBJECT:

At the termination of the ASTP mission, I would like to express my appreciation of the excellent support provided by all shifts, the ASTP team and maintenance personnel.

It was gratifying to see such close team work, sustained without friction between sections, during this comparatively long mission.

My personal thanks to everyone.

(R.J. SIMONS)
SE(O)

A letter from the Senior Engineer (Operations) (SOS) to members of the support personnel for the ASTP flight.

Apollo-Soyuz on display at the Smithsonian Air and Space Museum, Washington, USA.

NATIONAL AERONAUTICS AND SPACE ADMINISTRATION
GODDARD SPACE FLIGHT CENTER
GREENBELT, MARYLAND 20771

REPLY TO
ATTN OF:

September 1975

DEAR FELLOW MEMBER OF THE ASTP TEAM:

The successful completion of the ASTP mission has been achieved, and it is a sincere pleasure to forward your APOLLO/SOYUZ COMMEMORATIVE CACHET AND CERTIFICATE.

The cachet was designed to commemorate the NASA Tracking Network support for the Apollo/Soyuz Test Project, the first international manned space mission. It was postmarked as "first day of issue" at Kennedy Space Center, Florida on the day of the Apollo/Soyuz launches and commemorates the first official use of the twin Apollo/Soyuz United States postage stamps which were issued to honor this event.

These postage stamps are an international joint venture in themselves since the Soviet Union also issued twin pairs of stamps marking the joint mission similar in design but lettered with Russian. One of the twin stamps was designed by Robert McCall of Paradise Valley, California and the other by Anatoly M. Aksamit of the U.S.S.R. Such "first day covers" tend to become collector's items with the passage of time.

This cachet and certificate is presented to you to provide recognition of your outstanding contribution to the success of the Apollo/Soyuz mission. You have participated in and supported the world's first international manned space mission. This is an achievement of which you can be justly proud.

Please accept our sincere congratulations for the recognition you have earned as a member of the ASTP Team as exemplified by this award.

Tecwyn Roberts
Director of Networks

Above – letter accompanying the cachet and certificate shown on the facing page.

Lower right on facing page - The ASTP certificate awarded to the Author for his part in the flight. The medallion is made from an alloy of metal used in both spacecraft.

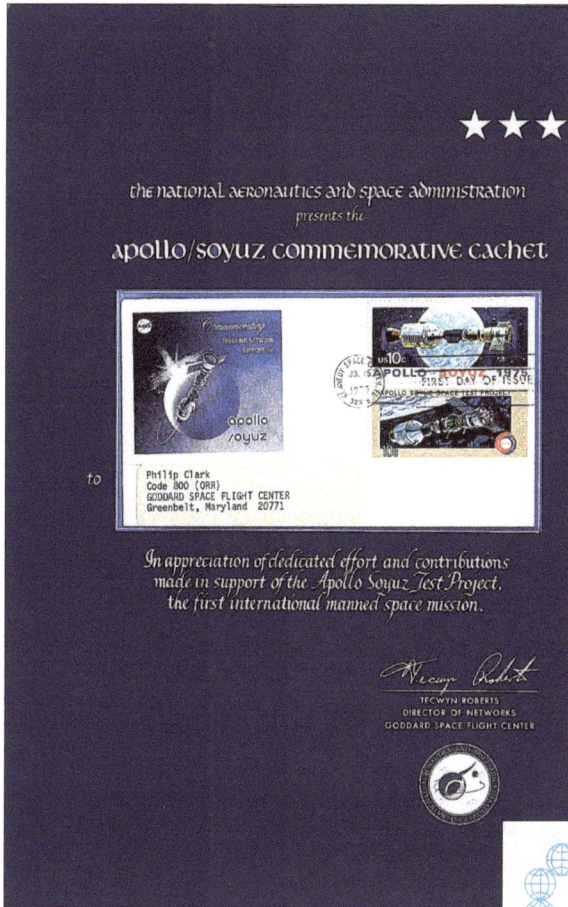

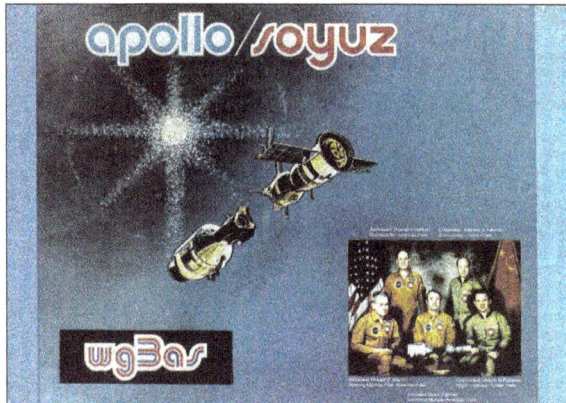

RADIO *VK2KPG*

SPECIAL EVENT STATION WG3AS IS PLEASED TO CONFIRM QSO AS FOLLOWS:

DATE	TIME (GMT)	REPORT	BAND		2 WAY
7/ /75			160 80 40 20		CW SSB
			15 10 6 2		AM FM
					OSCAR

WG3AS was licensed to the GODDARD AMATEUR RADIO CLUB by the Federal Communications Commission for operation during the flight of APOLLO/SOYUZ.

The Goddard Space Flight Center was named in honor of Dr. Robert H. Goddard who held the call letters 5ZJ. The center is located in Prince Georges County, Maryland, ten miles northeast of Washington, D.C., operates NASA's Global Tracking Network and performs basic space research. The Club is one of several employee recreational activities.

QSL via WA3NAN, Box 86, Greenbelt, Maryland, U.S.A. 20770

73

_____ OPR.

An amateur radio contact confirmation card (QSL card) issued by Goddard Amateur Radio Club to commemorate the ASTP Flight. VK2KPG was the author's callsign at the time of the flight.

Chapter Five – New Equipment for the Space Shuttle

Preparations for Orroral Valley to support the Space Transportation System, STS or Space Shuttle, began in 1977. The unique specialist equipment that had been previously installed for the 1975 ASTP support was removed, and new equipment for Space Shuttle was installed in its place.

Pat Lynch remarked on the equipment complexity:

"One of the other things I remember was installing the shuttle MUX. The voice/data multiplexer, which was the most incredibly complicated thing I had ever seen in my life, I think in concept. I met the guys who came in to put that into action but I can't remember much about them."

Considerable new equipment was installed to support the Space Shuttle. Sometimes things did not go as smoothly as planned.

Facilities Supervisor **Trevor Smith** remembered an incident:

"I guess the most memorable character at Orroral was the Chief Engineer, Dave Kemp. He was a memorable character because, as you recall, Orroral Valley had very narrow hallways and he was one of the few people we used to watch walking down the hallway, bouncing from wall to wall talking to himself. And sometimes you would see him around outside talking to the trees. He was a definite character. He was only a little bloke, but I saw him really arc up one day. One of our electricians had a bit of an issue with the logistics (stores) people. It was when we were coming up for shuttle support. There was a voice rack arriving from America and they needed it up and running the next day. Do you remember George McEwan? He always seemed to have some sort of a conflict with the logistics people for some reason. The result was that they would not give George the cable he asked for. So I went down with him to the store and they said no, they were not going to issue the cable. I then went around and saw Bernie Smith, he was the engineer responsible, and said to him sorry, we can't install your rack because they won't give us the cable. I said this because I knew what the implications would be. He said 'leave it with me', and he went and saw Dave Kemp. Next thing, Dave went around to the stores, and I have never heard a voice so loud come out of such a little person. We instantly got the cable!"

By far the biggest change was the installation of a new system for controlling the real-time flow of data received from spacecraft. This was the Digital Data Processing System or DDPS. This system was the start of a new concept for the handling of spacecraft data. Previously, almost all data received from spacecraft was in analogue form, and was recorded on analogue magnetic tapes.

These tapes were then shipped back to Goddard Space Flight Centre for processing, usually on a weekly basis. This involved considerable handling and shipping, and also meant a time lag between the data being received and then being available to the scientists who were waiting for the data from their spacecraft instruments. It was also a very costly method of getting the data. High quality analogue tapes were very expensive, and the cost of handling and shipping by air was very high. A method to reduce these costs was an advantage for NASA.

The new concept was that the later generation of spacecraft would transmit most of the data in a digital form. This would allow it to be processed immediately (in real time) on the station and then sent straight back over data lines. If for any reason the data could not be sent immediately it would be digitally recorded and then played back either later the same day or within a day or two at most. Thus, the manual handling and shipping of data tapes would be almost eliminated and the delays in getting data to scientists almost completely removed.

By 1977, many spacecraft were later generation, using digital data and only a few older spacecraft were using analogue data formats. The way of the future using digital data was to be less reliant on analogue magnetic tape recording with its attendant delays involved in shipment and handling. The use of real-time data, or near real time, meant that spacecraft controllers, engineers and experimenters could obtain data much more readily than waiting for the cumbersome analogue magnetic tapes. It also meant less handling of these heavy tapes on station.

The 642-B computer systems which had been used for manned space flight support were used to format the data onto three high speed data lines in a special sequence, so that the data could be decoded at the other end. These computers had been transferred to Orroral for support of Apollo-Soyuz as part of the station upgrade in 1974 and now were to be used with the Digital Data Processing System, or DDPS.

The new generation of spacecraft, including the Space Transport System (STS) - the Space Shuttle, would use the DDPS.

The DDPS system used three dedicated computers of the Digital Systems PDP-11 type. The system had a physically huge magnetic drum memory to use as a temporary store or buffer if data came in faster than it could be processed out. The drum memory was much like a computer hard drive, only it was about the size of a 50-litre petrol drum. To interface with this system there were a number of special operator control consoles. There was one for each receiving system on the station, and others for the data handling and the computer areas. From these consoles the telemetry data streams could be set up and controlled to go to each particular spacecraft control centre in real time, or recorded on computer controlled digital magnetic tapes for later transmission.

Each of the three main PDP-11 computers of the DDPS system was dedicated to a specific task. The Display Processor interfaced with the operator control consoles. It controlled the display screens at each console and processed the operator inputs. The Telemetry Processor accepted the data from the MSFTP-2, MSFTP-3 and the Dynatronics PCM-DHS data decommutators and, after formatting, processed it for passing to the File Processor. The File Processor controlled where the data was to be sent, either for real time transmission through the 642-B data transmission computer, or for temporary storage on digital magnetic tape. The File Processor controlled the drum memory and digital magnetic tape recorders and also took care of administrative tasks such as all of the data file designations and pass summary messages. An additional PDP-11 computer, which could be used for any DDPS function, was provided as a spare in case of a failure.

The DDPS system was installed in every STDN station around the world, and required specialised training for operation and maintenance. NASA decreed that each of the STDN stations should have a DDPS specialist and a training section was established for DDPS at the Goddard Space Flight Centre Network Test and Training Facility (NTTF). Each of the nominated station specialists was required to attend. The course was in two main sections - the DDPS File System and DDPS computer maintenance.

The nominated specialist for Orroral Valley was the author. The course was scheduled over a three-and-a-half-month period from 4 April 1977 to 24 June 1977 and the course certificates are seen below. During this time, he received a special certificate for 'unique performance' that was not awarded to anyone else – which is the bottom illustration below.

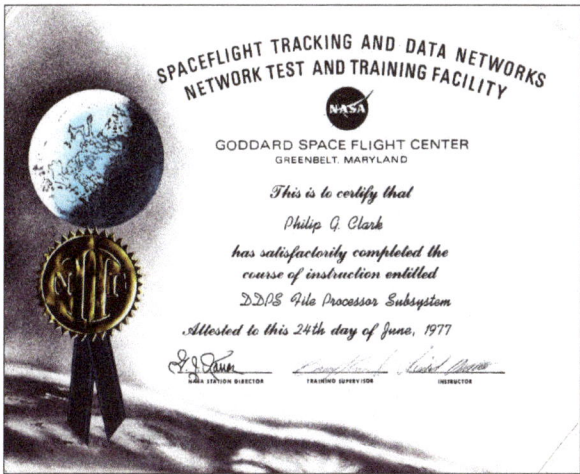

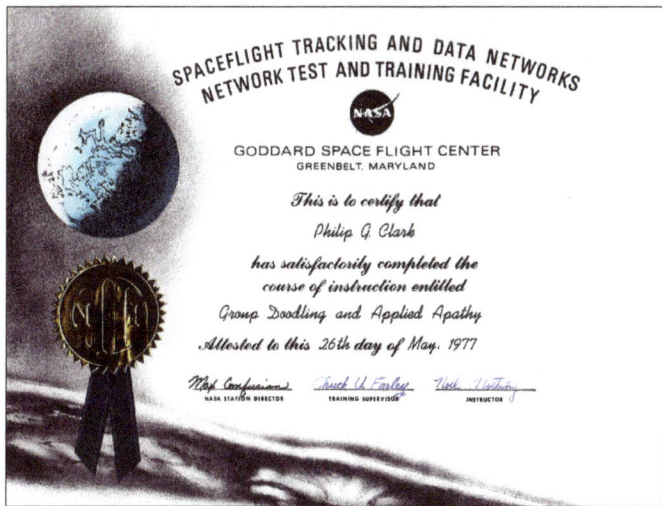

The DDPS remained the main data processing and data transmission system until Orroral station closed. In some limited respects the DDPS consoles for the telemetry links were similar to the old 'M&O' consoles of the MSFN stations. However, unlike the M&O consoles of the MSFN stations, special senior staff did not operate the DDPS consoles at Orroral. Some of the Operations Coordinators preferred to work from these positions, but often the consoles were operated by different shift staff, ranging in grade from a shift technician, to occasionally to the Shift Supervisor. In later years the DDPS consoles were even operated by some of the more senior equipment operators, further demonstrating the skills of these people.

The DDPS system also had the capability to receive command instructions remotely from the spacecraft control centres and then to send the commands to spacecraft via the SCE command systems when required. While the DDPS system was ideal for those satellites that used digital communication modes, it could not be used to support satellites that used analogue data, and there were still some of these being supported at Orroral. So the analogue magnetic tapes and other manual operations were still needed for these spacecraft.

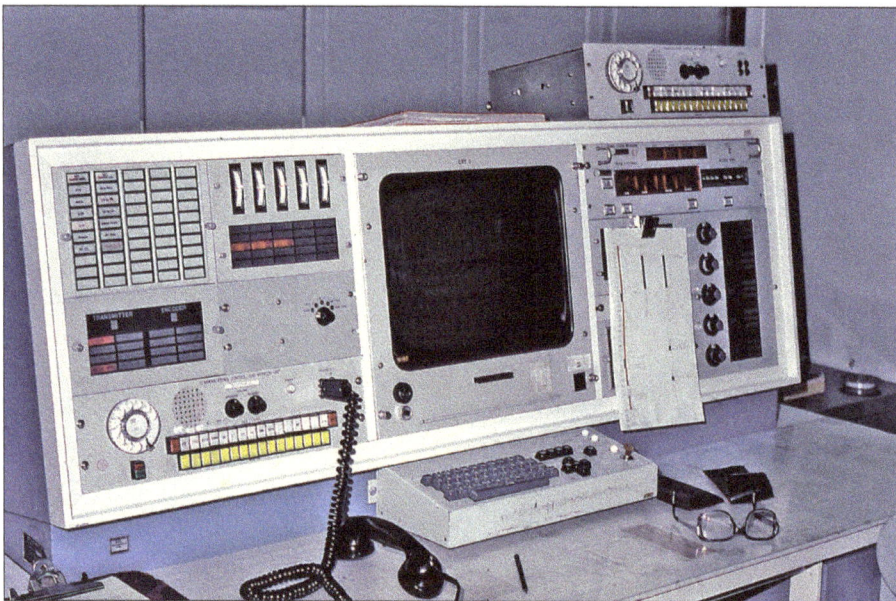

Top: One of the DDPS system control consoles. Bottom: The DDPS PDP-11 computers with the drum memory at far right.

Orroral Valley, Yarragadee Laser Station in Western Australia, and The Space Shuttle.

Unfortunately, the ground stations of the existing NASA tracking network could only cover less than 20% of the orbit of the Space Shuttle, even with augmentation from tracking ships and aircraft. NASA decided it was necessary to improve this coverage, which they did by installing a number of simple VHF voice communication stations around the world. These stations only had the capability of using radio for speaking to the shuttle crew. They had no capability to transmit commands or to receive any telemetry or television. One of these stations was installed at a location called Yarragadee on the west coast of Australia, south of the town of Geraldton. There was already a NASA facility at Yarragadee which was managed by Orroral. This was a laser tracking station.

Yarragadee was the name of a property on the road between the small port of Dongara, south of Geraldton in Western Australia, and the inland town of Mingenew. The laser site was situated on a leased section of this property, some 35 km inland from Dongara. The location is approximately 360 km north of the state capital city of Perth. A lease for the laser station site was signed in late 1978, the station was constructed in 1979, and became operational for laser tracking after September 1979.

The first Space Shuttle flight was due to take place in early 1981. As shown on the ground track map in chapter 8, the first orbit of the flight would pass over both Yarragadee and Orroral tracking stations. Both stations had to be fully operational and ready to support this first critical Space Shuttle mission.

In the second half of 1980, a transportable air-ground communication station was sent to Yarragadee. This consisted of an antenna together with a van which contained the controls to operate the antenna, and the transmitting and receiving equipment for the air ground voice communications. A small tower had been installed at the laser site on which to mount the antenna. The antenna was shipped dismantled in several crates.

A team, consisting of the author and two other technicians, was tasked with the job of installing the antenna, interfacing it to the equipment at Yarragadee, and commissioning the system. In January of 1981 the team flew to Perth and then drove north to the site to install the equipment. On arrival they found that the equipment had arrived and was still crated in its boxes. A large mobile crane stood by ready to install the antenna system. With the assistance of the Yarragadee station staff, the equipment was un-crated and the antenna assembled ready for installation. So far things had been going well, a sure sign that something was about to go wrong.

The Air-Ground voice equipment van is shown with the cables connecting it to the antenna.

Above and below: The Air-Ground voice antenna at Yarragadee.

When sufficient assembly had taken place, it was time to install the antenna pedestal on the tower. It was a beautiful summer's day with the sun shining brightly from a crystal clear brilliant blue sky. The author climbed up the tower and had a look at the mounting ring where the pedestal would be bolted. Something did not look quite right. He checked the documents and had another look at the mounting ring on the tower, it still did not look quite right. He recalls vividly the conversation with the station supervisor, Keith Harris:

"Keith" I said, "The mounting ring does not seem to be in the correct place. When they installed it, did they survey it accurately?"

He gave me a slightly puzzled look and said, "Survey? I don't remember seeing any surveyors."

I said, "Well how did they line up the west alignment bolt hole for the mounting ring on the tower?"

He said "All I remember is that they welded the ring on the tower top and then put it up and just seemed to line the whole thing up with the fence."

"Oh!" I said.

The author continues:

"At this point the crew and the crane driver were all looking at me expectantly. The crane had driven the distance of some 360 Km from Perth to the site and needed to get back as quickly as possible. It was nearly noon and the crane bill was ticking. I looked at the Sun in the sky and checked my watch. I had no compass, I had no surveying equipment, so the best I could do in a hurry was using the sun and my watch work out approximately where west should be. Using this rough figure, I arbitrarily assigned one of the bolt holes on the mounting ring as being nearest west, and directed the crane to lower the pedestal onto the tower so that it could be bolted into place.

This was done, and the crew then bolted the antenna pedestal into place. Next, the assembled antenna array was hitched to the crane so that it could be raised and installed on the pedestal. The crane was then allowed to depart for Perth, leaving us with our problem growing darker with the setting sun. While the crew were installing the antenna system and the interface cables back to the equipment van, I was left with the problem of working out how we could ensure that the antenna would point where it was supposed to. The station had no compass and no surveying equipment. As I puzzled over the problem, I realised that here was a laser tracking station. It was very accurately surveyed. Also, it had a calibration site board, or target, some distance away on a slight rise which was very accurately surveyed, and the angles and distance to which were accurately known.

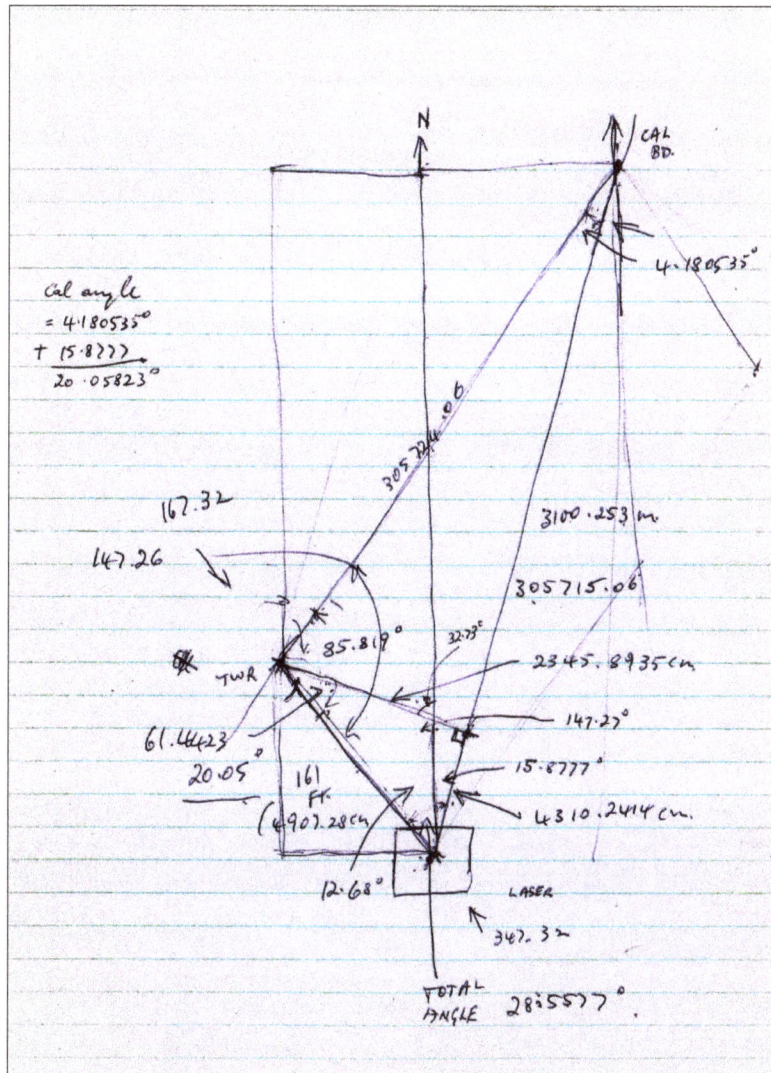

From this I deduced that I had two reference points. I did have my pocket calculator with me, which was able to do trigonometry functions. I borrowed a tape measure from the supervisor and carefully measured some distances on the ground from the laser to the centre of the antenna tower. I drew a line on the ground from the laser along the bearing of the calibration target, and then made some measurements. Using these measurements and the known angle for the laser to point at its calibration target, I used trigonometry to work out the angle that the antenna would need, to point at the same target. Having worked this out I positioned the antenna as low as possible, climbed onto the structure and sighted along the antenna booms to get it pointed at the laser calibration target. When this was done, I directed one of our team members to go to the equipment van where the antenna angle indicators were, and another to go up to the antenna and loosen off the servo encoder that generated the antenna angles for the readout. I told the person up on the antenna to rotate the servo angle encoder until the operator in the equipment van saw the angles that matched my calculation. Everything was then tightened up again and to the best of my knowledge, that is how the antenna remained to track the space shuttle. A copy of my calculations is shown beside this text.

However, I was faced with another not so small dilemma. Once all of the equipment was installed there was nothing to test the antenna or the associated transmitters and receivers. No test equipment had been sent to the site. I said to our team that we could not leave without at least ensuring that the equipment was working. I thought about this problem for a while and then realised that I had some knowledge which might be useful. This knowledge came from my radio background, and also the fact that I was a licensed pilot. I knew that the space shuttle air-ground voice communications utilised the VHF military air band using amplitude modulation. I also knew that there was an Australian Air Force Base at Pearce, near Perth, about 360 km to our south.

I reasoned that because we had a very sensitive antenna and top-quality equipment, there was a good chance that we would be able to hear some communications from any airforce aircraft approaching the Pearce Air Force Base. I found the appropriate frequencies and tuned the receivers. We sat listening for some time, and eventually heard some voices. I said to the team that we now knew that our antenna and receivers were working. But how to test the transmitter? I made a couple of telephone calls and said to the team that we could test the transmitters, but that we would only get one shot. I set the transmitters up appropriately, and the next time we heard voices I keyed the transmitter and spoke a few words. Immediately I received a surprised reply. I said to the team that we now knew the transmitters were working, but we must not do that again! I said I had been talking to the pilot of an airforce aircraft on approach to Pearce Air Force Base. At least now I could leave the station with some confidence that the whole system was working. However, the story was not yet over. I was not to know, at this point that, had I used the frequencies designated as prime for space shuttle support, it is unlikely that this test would have worked."

A short while after the author returned to Orroral, NASA sent a test kit to each of the new air-ground stations to test the antenna and its equipment on site. Part of this kit was a test procedure to be performed at each station and the results sent to GSFC, Houston, and all other stations around the network. The results from Yarragadee came to the author at Orroral. When he perused them, he noted that something appeared to be not quite right.

He continues:

"I called Yarragadee and said that there appeared to be a discrepancy, and would they run the test again. They did and the results were the same."

Within a few days, results started to come in from the other air ground stations around the world. However not all of these stations had the same type of antenna system as that used at Yarragadee. From memory, I think that there were only four with the type of antenna which was used at Yarragadee. I noted that the results from all of the stations with this type of antenna had the same discrepancy. The impact of this discrepancy was that, in effect, on the prime frequency designated for the space shuttle voice communications, the antenna did not work any better than a piece of wire. This was a serious fault.

I took the results to Dave Kemp, the Chief Engineer at Orroral, and discussed the matter with him. He asked if I had verified our figures and I said we had. He said that to be on the safe side we needed to run the tests again. So I contacted Yarragadee and asked them to run the tests again. The results were identical. I discussed the test with Dave and he considered the matter to be serious enough to inform mission control at Houston. He also considered that because the matter was important and reflected on Orroral it needed to go through the Station Director. When he was informed, the Station Director again queried the figures and the Chief Engineer assured him that the figures were correct. It was then decided to tell mission control at Houston that, based on the test at Yarragadee and certain other stations, it appeared that these particular antennas may not work correctly for the space shuttle mission as planned. This caused considerable consternation. It appeared that these antennas may have had this fault since manufacture.

The concern from Houston was that the frequencies to be used by the space shuttle had been negotiated for worldwide coverage and they could not be changed easily, or possibly even at all, in the timeframe available before the launch of the spacecraft. I then suggested that, because the antennas worked correctly at the backup frequency, all that needed to be done was to swap the prime and backup frequencies for the mission. There were expressions of relief at such a simple solution. The mission documentation for the first space shuttle was changed to reflect the swapping of frequencies and I believe that this is the way all subsequent space shuttle missions which used ground stations for voice communication, were supported. That was part of my contribution to the Space Shuttle missions.

Chapter Six - Training for the Space Shuttle

The first Space Shuttle flight took place in 1981. Training at Orroral began in the last quarter of 1980. Tracking the Space Shuttle was the most complex and demanding operation Orroral ever experienced. While the station used the now normal tracking equipment to support the Space Shuttle mission, there were new and specialised procedures to work with.

Orroral had not tracked a manned spacecraft since Apollo-Soyuz in 1975.

Most of the procedures used for tracking Apollo-Soyuz were not applicable to the Space Shuttle and there had been many staff changes in the tracking shifts in the intervening six years. Considerable training was required for all operations staff to work with the different techniques for Space Shuttle tracking operations. Much of this was done by practice through a series of simulated passes, during which the station engineers created 'emergencies' that might occur.

The training for the Space Shuttle was quite different from Orroral's previous manned tracking, ASTP. In the case of ASTP there was new equipment as well as new procedures to come to grips with. However, for the Space Shuttle there was virtually no new equipment.

The last major equipment upgrade, the Digital Data Processing System (DDPS), had been in use since 1977 and staff had had time to become familiar with its operation. Further, the Space Shuttle would use the other standard equipment which was being used every day for many other spacecraft.

What was needed in the training was for staff to learn the particular operational procedures for Space Shuttle, and what to do when any piece of equipment failed. This training was simply a matter of repeated practice by carrying out simulated passes.

Richard (Dick) Elliott recalls the training:

"The mentally draining part of that was in the simulations. Later, you never had as much go wrong on a real pass as you did in the simulations. You were keyed up ready to pounce on anything that went wrong. In some ways the passes were almost uneventful by and large, in that nothing much went wrong."

The specialist areas designated for Space Shuttle operations were:

- Operations Supervisor,
- 642B computer operations,
- Data handling and,
- Air to ground voice.

This last position was the connection for voice communications between the ground (Houston Mission Control Centre) and the Space Shuttle astronauts. The author was the senior technician in charge of the Air-Ground communications for Orroral and thus Australia.

For the first flight of the Space Shuttle, Orroral Valley Space Tracking Station put together a special team of experienced staff, the so-called shuttle 'Tiger Team'. The 'Tiger Team' provided staff for the specialist areas and took part in every training exercise which was conducted. They tried to work with each of the normal tracking shifts so that as many people as possible had some hands-on practice with Space Shuttle operations.

Unfortunately, there was one position that could not be easily simulated for the Space Shuttle. This was the antenna operator, or driver as they were commonly known, for the dish antennas. In the period of time from ASTP in 1975 to Space Shuttle in 1981, antenna drivers had gained much experience tracking other space craft on S-Band frequencies, but until a shuttle was launched, and was in the correct orbit, the exact conditions could not be duplicated.

Of course, the antenna driver was crucial to the success of tracking any pass. If the antenna did not acquire the signal, then nothing else mattered. Although other spacecraft did not provide exactly the same signals as would be received from a Space Shuttle, this practice for the antenna drivers did sharpen their skills.

Peter Uzzell -

"The Shuttle operations were pretty significant. I went to Goddard and did the shuttle course there. Then the first shuttle operation was WOW! but then became a bit ho-hum like the rest of it did eventually."

One of the most significant successes of Orroral support for the space shuttle was the appointment of Peter Uzzell as the leader of the specialist support team, the 'Tiger Team'. Peter had served with the Australian Army before coming to Orroral He was initially a technician grade C. His first operations role was as an Operations Coordinator in Operations Room number one, with the 26-metre telemetry system. After serving there for a time, he was appointed as the Shift Supervisor of C shift, in which position he eventually became the longest serving supervisor of that shift. Peter was well liked, seen as very competent, and was a good supervisor without being pedantic. He was well respected by the personnel who worked with him and very few had any criticism of his style. In 1981 he was promoted as the Senior Operations Supervisor (SOS).

His selection of personnel and guidance during the training and later tracking of was, in the opinion of the author, one of the major factors for the superb performance of the Orroral Tracking Station in tracking the Space Shuttle. This laid the foundation for a continuing excellent performance at Orroral in tracking the Space Shuttle until the closure of the Station.

Darryl Fallow -

"Part of the success of the people on shuttle support was good leadership. I am thinking particularly here of the time when we were leading up to and during the space shuttles. I think Peter Uzzell was largely responsible for that. The way that things worked was they had a separate tiger team so we could concentrate on the day to day stuff and others could concentrate on other things when things got difficult, that approach worked quite well."

The first 'Tiger Team' was very comprehensive to cover almost any possible contingency which might have occurred. The tracking and operations part of the team were:

Peter Uzzell	OPSR
Philip Clark	Air/Ground Voice
Basilio Ormeno	Data Handling Supervisor
Ken Strickland	642B Computers

In case of equipment failures or other problems, additional support and maintenance staff were assigned to the team for the first flight. They were:

Heinz Assell	Lee Hopson	Dave Barter
Graham Johnson	Sid Benning	Mike Lampard
Bill Brooke	Tony McEvilly	Kevin Carroll
Peter Militch	Len Cowling	Ross Murray
	Ian Edgar	

Basilio (Bas) Ormeno -

"During the training everything seemed to go so smoothly that we did not have any troubles at all. Although the Shuttle was not due until 1981, we started training in 1980. One of the things the Americans brought out for us was a transponder they put on a plane for us to track."

Darryl Fallow -

"The training was very rigorous and seemed to go on interminably! There was a lot of training and it was very thorough. I think the reason it dragged out was because of delays in the Shuttle launch. An American team brought a transponder for us to try and track. They put this thing, I think from memory, on a helicopter that floated around pretending it was a Shuttle. We had all sorts of fun trying to track it because the antennas would not track it properly. This was probably because of its proximity to the ground. Radio reflections from the mountains stuffed up the antenna because it was so sensitive and so it did not know where to point. The net result of these attempts left people a little bit jittery when the actual time came to track the Shuttle."

Ken Strickland -

"I remember the training as being pretty comprehensive and the fact they had people standing around making things break. I was on the 642B and I sat around waiting for them to pull the plug on things, which they did!"

Chapter Seven - The Space Shuttle

The concept of a reusable launch vehicle began in the 1960s. Several ideas were proposed, some of which could almost have come from science fiction. The motive behind this was to try to save some of the expensive material that was simply discarded and lost with the large launch vehicles which were being used prior to this. The first concrete developments going toward a reusable launch vehicle came in early 1969 when NASA let contracts for feasibility studies of reusable vehicles. A number of companies produced proposals for study, and NASA set up a task group at their headquarters to look at the various concepts which had been proposed. All of this was being done as the first men flew to the moon. It was in 1970 that NASA proposed a more firm basis for building a reusable launch vehicle and awarded contracts for the study of these. One of the first concepts looked very much like the space shuttle as it was eventually launched, however there were many iterations and alterations before it was finally developed into the reusable space shuttle that finally flew into space.

To put this spacecraft into context, at the time that the USA's first satellite was launched, the most powerful rocket available was only able to put a payload of about 28 Kg into orbit. Originally the Shuttle was designed to carry a payload of just over 11,000 Kg into orbit. However, representations from the Department of Defence had this payload increased to 29,500 Kg in an equatorial orbit or 18,100 Kg into a polar orbit. (The difference is due to the speed assist given by the rotation of the Earth when launching into equatorial orbit.)

To place something into orbit is more complicated than just shooting a rocket straight up. If this is done, when the fuel runs out it will simply fall back to earth. In order to get into orbit, not only does the rocket have to go up, but it also has to roll over a little and accelerate to the correct speed for orbital injection as well. While doing this, it still has to climb above the atmosphere to the orbital height. If you listen to a recording of the commentary of a space shuttle launch, you will hear them refer to a 'Roll Manoeuvre'. This is to accelerate as well as climb. This type of launching takes much more energy than simply going straight up.

In about mid-1972 the shuttle concept was finalised. The delta-wing Space Shuttle would have new engines using liquid hydrogen (LH2) and liquid oxygen (LOX) as fuel. This fuel would be carried in an external tank which would be jettisoned just before orbit. The tank contained 101,000 Kg of hydrogen and 603,000 Kg of oxygen. (Oxygen atoms weigh 16 times more than hydrogen atoms.) This huge amount of fuel provided only about eight minutes of power for the shuttle main engines. The whole tank full of fuel was over one third of the lift-off weight of the shuttle, which was about 1,998,000 Kg. Attached to the external tank were two solid-rocket boosters that were jettisoned about two minutes after launch and dropped off. They fell back to the sea on parachutes for recovery and later were refurbished for another flight. The solid boosters were expected to last for 20 flights. The shuttle itself was expected to have a life of about 100 flights.

The electronics for the shuttle at the time of design and development were state of the art. However, by the standards of 2016, when this book was written, they appear quite antiquated. For example, each of the five digital computers on board had a capacity of only 48 K words. This was the memory size of the 1970s Apple computer. It was minuscule by the standards of the 2000s when computer memories had reached gigabyte size.

The guidance systems were inertial navigation systems. They were similar to the type as used in the early 747 airliners. Such a system could take an airliner from Sydney to San Francisco and put it in the correct parking bay, however compared to later GPS navigation, they were not as reliable nor accurate. But that was the technology of the time.

The Shuttle could carry up to seven crew. Three of these were the actual flight crew, and up to four were payload specialists.

The following facts give some idea of the enormous power required to lift the Shuttle and its payload into orbit.

At 3.8 seconds before lift-off, the Shuttle main engines were started while the launch vehicle was kept clamped to the pad. Because these engines are not centred on the launch vehicle, the Shuttle actually bent backwards toward the external tank by about one metre. This was known as the 'twang'. While this happened, the main engines built up to 90% power.

Then there was a 2.64 second delay to allow the Shuttle to 'twang' back to the vertical position, at which point the solid rocket boosters were fired, the clamps released, and the Shuttle lifted from the pad.

After just 50 seconds from launch, the vehicle passed the speed of sound. The solid boosters only lasted for 2 minutes and 12 seconds and were jettisoned, but by this point the shuttle was travelling at 4.5 times the speed of sound and was 45 kilometres high! The Shuttle continued on with its main engines and by six and a half minutes it was travelling at 15 times the speed of sound and was 130 Km high. Here there was a special manoeuvre to jettison the external tank. The Orbital Manoeuvring System (OMS) engines were then fired to place the Shuttle into its initial orbit. To get to this point had taken only about 10 minutes, and the shuttle was travelling at about 28,000 Km per hour! About 45 minutes after launch, the OMS was fired again to place the Shuttle into its correct orbit,

The Space Shuttle program ended with the landing of the Space Shuttle 'Atlantis' on 21 July 2011. STS 135 was the final flight of the program. Although original planning thought that the Space Shuttle would fly about 45 flights per year, from the mid-1980s, by the close of the program in 2011 a total of only 135 flights had been made.

Summary of Shuttle Flights

Shuttle	Orbiter no.	Flights	Orbits	First flight	Last flight
Columbia Note 1	OV-102	28	4,808	STS-1 12/4/1981	STS-107 16/1/2003
Challenger Note 2	OV-099	10	995	STS-6 4/4/1983	STS-51L 28/1/1986
Discovery	OV-103	39	5,830	STS-41D 30/8/1984	STS-133 24/2/2011
Atlantis	OV-104	33	4,848	STS-51-J 3/10/1985	STS-135 8/7/2011
Endeavour	OV-105	25	4,677	STS-49 7/5/1992	STS-134 16/5/2011

Note 1: Columbia was destroyed during re-entry on 1/2/2003.

Note 2: Challenger was destroyed during launch on 28/1/1986.

Chapter Eight – Tracking the Space Shuttle

The first four flights of the Space Shuttle were development flights (DF) (sometimes called research and development flights), to test the concept and systems of the Space Shuttle prior to any operational flights carrying working payloads and additional specialist crew. The only crew on the development flights were a flight crew of two, the commander and the pilot. The development flights carried an extra radio downlink for development flight information data, the DFI data downlink. Normal flight data was carried on the operational data downlink, the OD downlink. The on-board tape recorder data was 'dumped' (downlinked) at much higher data rates on the FM downlink when in view of a tracking station.

The first flight of the Space Shuttle to actually achieve Earth orbit launched on 12 April 1981 for a two-day six-hour flight. During this flight, Australia was honoured by the crew of 'Columbia'. While over Australia during orbit 16 they played a short excerpt of Slim Dusty singing 'Waltzing Matilda'. To the Author's knowledge, such an honour was never given to any other country during the Space Shuttle Flights.

The first operational flight of the Space Shuttle was later also flown by 'Columbia', launching on 11 November 1982 for a five-day two-hour flight. As well as the flight crew, it carried two mission specialists. It was the first flight to deploy a satellite (PAM-D or SBS-C).

Tracking the Space Shuttle was something of a stop-and-go process. This was caused by three main factors. These were; the period or time of each orbit, the angle of the orbit with the equator (inclination), and the rotation of the Earth. What these factors did was cause the tracking passes to be in view of the tracking station in groups. The tracking passes themselves were about 90 minutes apart, because of the orbital period of the spacecraft. Then there would be a longer break as the Earth rotated to bring the tracking passes in view again. For some of the Space Shuttle flights, this longer break was between 12 to 16 hours, depending on the orbit. Groups of tracking passes could be up to as many as five, or as little as two or three, depending on the orbit of the spacecraft. Usually, when the passes were in smaller groups, the break between groups was shorter. This meant that sometimes it was necessary to have two specialist teams to allow tracking staff to have adequate rest periods.

Before each group of tracking passes, the station systems and equipment went through a three-hour System Readiness Test or SRT. What was done during this test and check was set out in a countdown document timed to when the spacecraft was expected to appear over the station horizon. The final interface test with Goddard Space Flight Centre and Mission Control at Houston was scheduled to take 45 minutes and so was called a Horizon minus 45-minute (H-45) sequence. The H-45 sequence table is shown in Appendix 6 together with a transcript of the actual voice communications that took place at Orroral during that process for one group of passes during the first flight of the Space Shuttle in April of 1981.

Before a Space Shuttle pass or track, the equipment needed to be configured, just as for any other pass or track. However, in the case of the Shuttle, there was just a little more because of the number of telemetry radio links and data requirements. The Local Operating Procedure document that detailed the complete setup for Orroral equipment was about 190 pages. (See Appendix 5)

SHIFT SCHEDULE FOR 83.0406

	LINK – 1				LINK – 2
1530				1531	J1295 06912
1545				1540	R C 2W WB PDF-A,B
					SS 1506-1542
1600				1600	M2006RT STS-6
1615					SRT FOR MINI SIM
1630					
1645					
1700					
1715					
1730				1730	M2006SM STS-6
1745					R C WB(168K)
1800					MINI SIM
					BM 28/1958
1815				1815	ORBIT 001
1830				1823	SCM 05/2032
1845				1850	ORBIT 002
1900				1858	SCM 05/2032
1915					
1930				1925	ORBIT 003
1945				1933	SCM 05/2033
2000				2000	
2000			2000	2000	
2015	2000	A1096 00796	2000	2000	A1120 15969
	2015	C(6M)2015-2035	2015		P WB PDF-J 100K
2030	2015	R WB PDF-F STRD	2015 CSAD(6M)2015-		
		DUAL SUPPORT	-2105		
2045		SS 2000-0221	R WB PDF-D STRD		
			DUAL SUPPORT		
2100			SS 2000-2225		
			2050 C 2050-2105		
2115					
2130				2120	
2145					
2200					
2215				2209	
2230	2223		2223	2229	G1291 01827
2245	2224				R C WB PDF-A,B
	2242	A1037CS 40525			TR DUMP
2300	2258	R C(6M)WB STRD			PWI MODE-A 2239-0047
		PDF-A,E			SS 2159-0049
2315	2300	PBM 05/1918			
		SS 2224-2300			
2330	2315	A1096 00796			
		R WB PDF-F STRD			
2345	2338	SS 2300-0221			
		CS & C(6M)2338-			
0000		-0038	0000		
0015					
0030				PROGDF ET	
0045	0219		0100	RC48U, WB 50K	
				IUS SOFTWARE	
				SEQ 4006D &	
				4006MD	
				BM 04/1655	0047

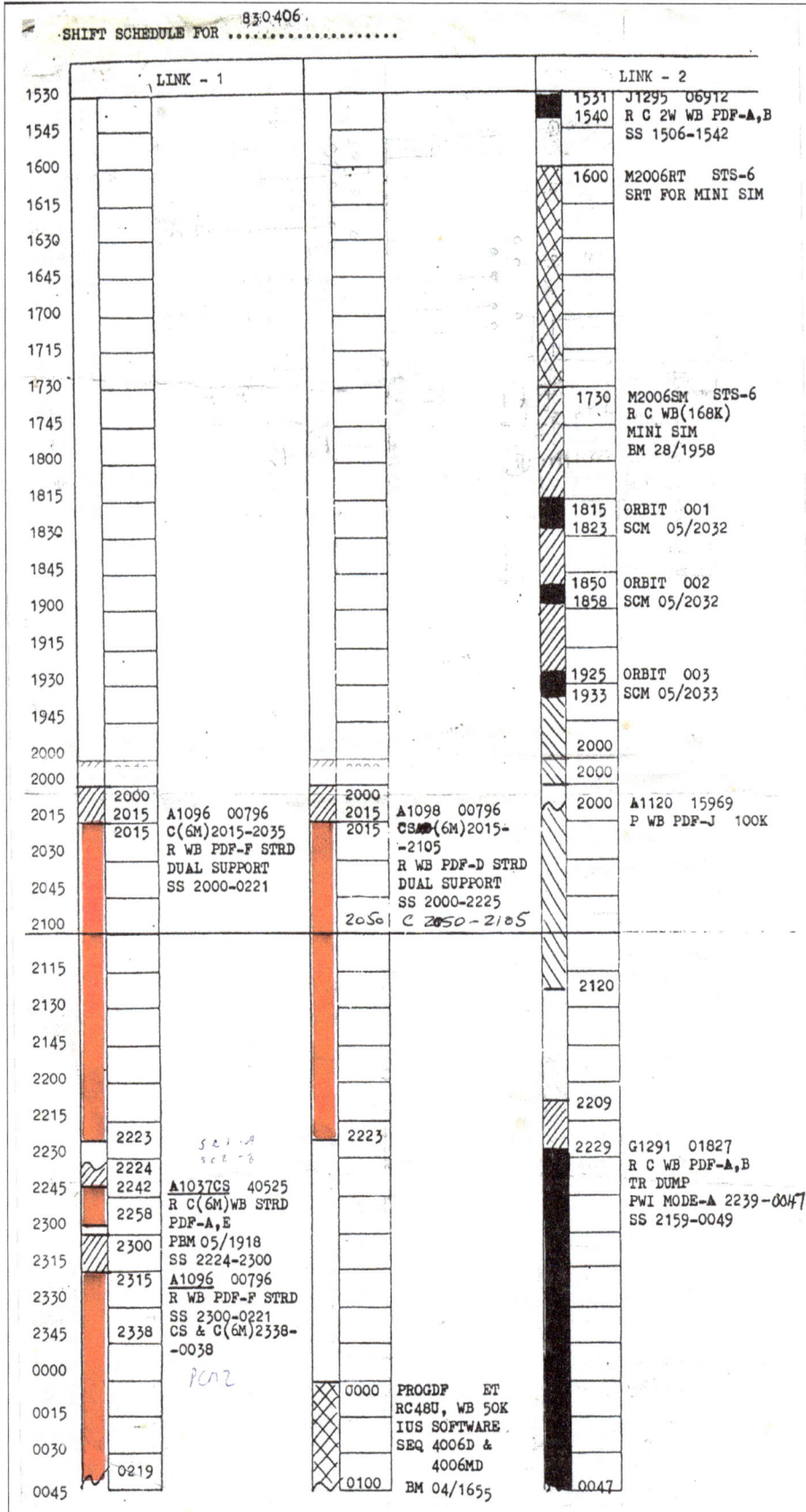

Part of the daily tracking Schedule at Orroral for 6 April 1983.

This was about twice the size of LOPs for some other complicated spacecraft and was large compared to LOPs that were only about 12 pages for uncomplicated spacecraft setups.

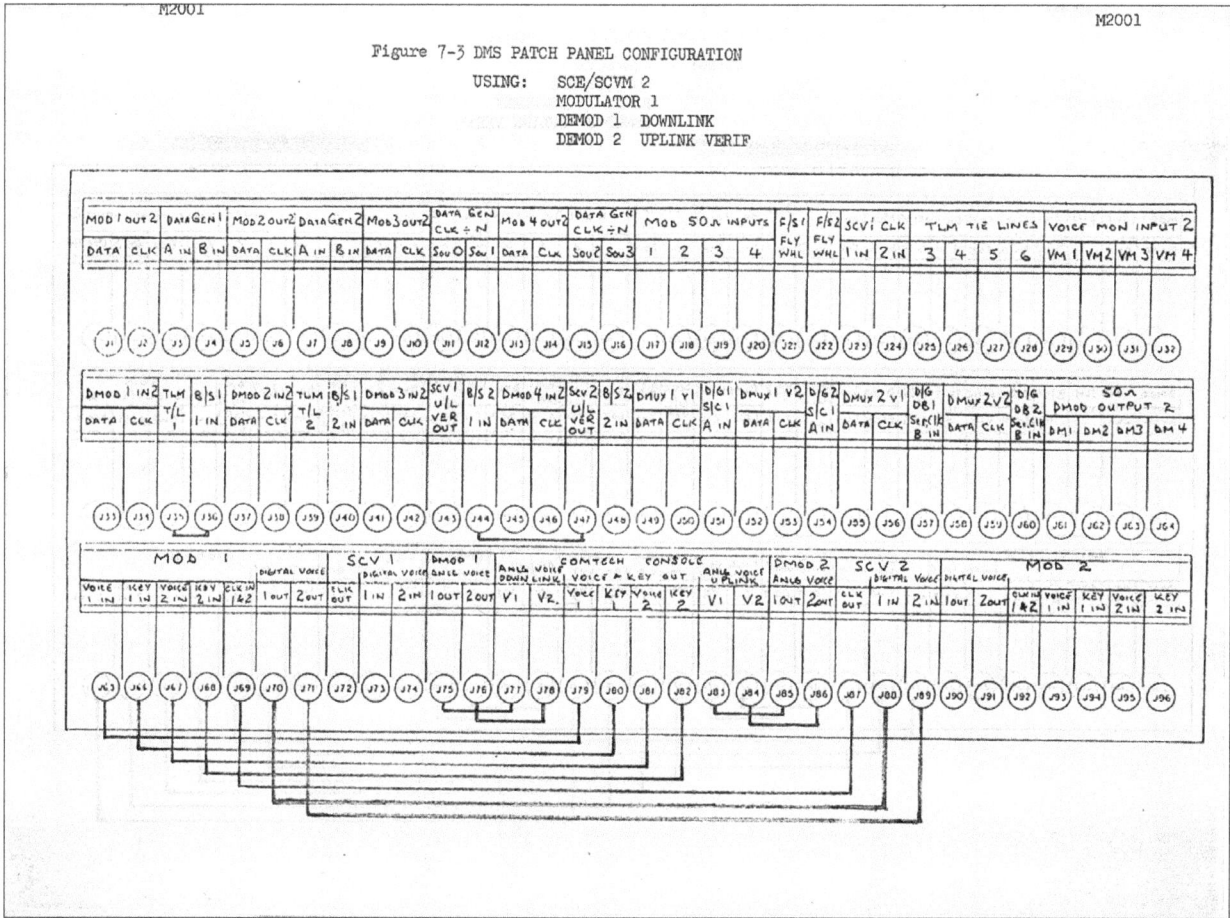

Figure 7-3 DMS PATCH PANEL CONFIGURATION
USING: SCE/SCVM 2
        MODULATOR 1
        DEMOD 1 DOWNLINK
        DEMOD 2 UPLINK VERIF

A diagram from the LOP of one of the coaxial patch panels. This patching is for one of eight possible different equipment configurations of the command equipment that could be used.

There was a three-hour period allowed for setup and configuring of the equipment and systems before the H-45 interface to Goddard Space Flight Centre and Houston Mission Control This was the System Readiness Test or SRT, as described earlier.

When an SRT was required, it was shown on the daily tracking schedule for the station. An example can be seen in the tracking schedule pictured here, at time 1600, as 'M2006 STS-6 SRT for Mini Sim'. The H-45 Interface period is seen on the schedule shown at time 1730. Although this particular example is for a practice simulated pass only, the scheduled times would have been shown in the same way for the actual tracking pass. The receivers, data recorders, data handling (processing) equipment, computers, command and voice systems could all be set up at the same time. This was all done during the SRT period in accordance with the LOP. The antenna operator had positioned the antenna to point at the collimation tower and had carried out tracking tests. A technician had configured several signal generators into the collimation tower for receiver testing and setup. These signal generators were also used for testing the data flow through the entire system when it was configured.

For the first four shuttle flights (development flights) there were nine receivers to be set up. Two for main data (Operational Data, OD) and two for main data backups, one and a backup for Development Flight Instrumentation data (DFI), two for FM television and on-board tape recorder data (DUMP/TV), and one for acquisition aid use.

Six data tape recorders had to be set up. Three of these would run for the pass to record the data and tape dump or television with the other three as backups. The recorders were loaded with large 14-inch diameter reels of half-inch wide tape 7,200 feet long. At the recording speed of 120 inches per second these large reels only lasted for 12 minutes. (The tape was passing the recording heads at 11 Km per hour!)

Although the most important data was sent to Houston MCC in real-time, there was other data, such as on-board tape recorder dumps at very high data rates, which could not be sent immediately. These were recorded and played back at reduced speed after the tracking period had ended. In the data handling area, a more complex setup than used for most other spacecraft was needed. This included three Manned Spaceflight Telemetry processors (type MSFTP-2 and MSFTP-3) plus frame synchronisers, PSK demodulators, and word formatters.

Five Multifunction Receivers in the link 2 area of Orroral's operations room.

Two Manned Space Flight Telemetry Processors type 2 (MSFTP-2).

The 642B computer technician had set up the computers for data transmission and to receive commands. The DDPS system was already configured to support other spacecraft at the same time as the shuttle, but the shuttle configuration also needed to be set separately in this system. The command and air-ground voice systems also needed to be set up and tested. Once all equipment was configured, testing of the complete system was carried out.

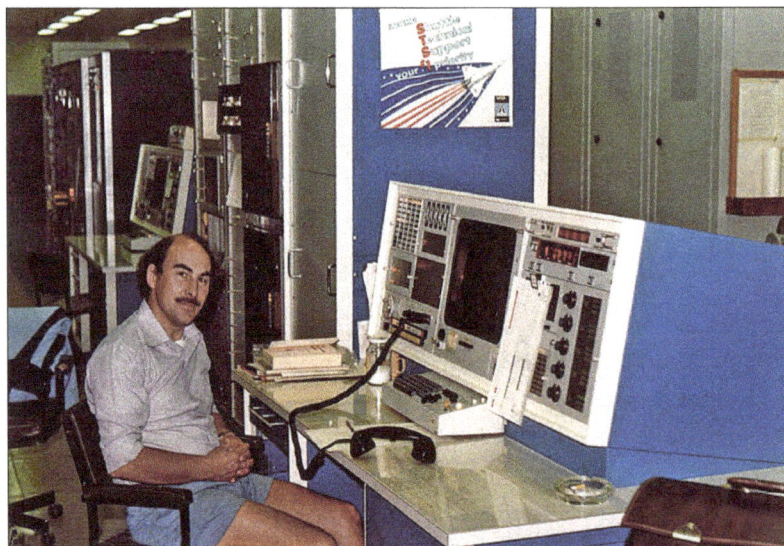

Rob Quick at a DDPS operations console

The Air-Ground voice control console

The Author operating the console as shown on Canberra ABC television news..

The uninitiated reader of the Local Operations Procedure may have thought that the complicated setup required would have resulted in a frenzy of activity, but such was not the case. The setup was accomplished by technicians and operators working with systems they knew well and had practised with often.

Pat Lynch gave his impression of how the tracking operations were set up:

"It all appeared to me as if they went about it with calm efficiency. There were very seldom panics. There were occasional panics, obviously, if something failed at an inopportune time, but generally it was just a matter of guys talking among themselves, often with a fag hanging out of their mouths. You were still allowed to smoke in those days. They were just grabbing a bunch of patch cords and just seemed to do it casually while they carried on a conversation. There wasn't a panic that I had observed quite often at other overseas stations I had visited. But I thought these guys really seemed to know what they're doing. And that was the crux of it. I think it was because this happens so frequently and it was just a routine task."

Orroral was one of the four stations around the globe to be have the capability for real-time television to be received. The other stations were Merritt Island in the USA, Madrid in Spain, and Goldstone in the USA. For the first orbital flight of the Space Shuttle, Orroral saw the passes in groups of between three and five in sequence. This first Space Shuttle flight lasted for just over 54 hours or 36 orbits. Before each group of passes there was a three-hour set-up period to configure the equipment and systems. This was the Systems Readiness Test (SRT) period.

One of the technicians, Ernie Cook (Cookie) had some influence over this test.

Ross Murray recalls:

"Ernie Cook used to spend most of his day sitting there writing software for the 642B's. In the early days, when the shuttles first started, the program they had to run took 2 hours to do a confidence check on all of the gear. It did all sorts of tests on the magnetic tape units and on the 642B computer itself. The trouble was, the shuttles were coming over about every 90 minutes and they were supposed to do a confidence check before they did each pass.

The first one was all right, but from then on there was no way you could actually do the confidence check, because the shuttle was back again. Ernie wrote this diagnostic program which was brilliant! It did the whole test of the gear in fifteen minutes. When some of the Yanks coming out for the early simulations saw this, and what's more, the fact that the operator didn't even have to get out of his chair, they said Hey! Now these guys came from Houston, not from Goddard, and Houston had a whole different sort of attitude to what Goddard had. So they looked at Cookie's software and said hey that's a great program. They said to Ernie, look you've got to submit that. Ernie said it's too much trouble, all that paper work, you've got to give commented listing and the whole deal. But in the end, he got talked into it. So off he went and did all this.

Well a little while later, back comes the program all official, no problem, from Goddard and credited to code 802.3 at Goddard. One day we were sitting there and Dave Kemp came in and he says to Ernie, can I see you in the office for a while. That was a bit unusual. Ernie was quiet, and could be quite sarcastic. So, Ernie goes into the office. Derek Trushell, the digital engineer, was not there at the time. There was a cable tray was built into the floor for cable to the front offices. The cable tray ran around the outside edge of the room to run cables into Dave's office. So we all went racing over to where Derek had his chair normally, we lifted off the lid of the cable tray, and listened down the duct. Ernie walks in and Dave says listen, Goddard has been pushed by Houston into all of this and there is a special commendation they have sent out for you. Here is this special letter. You could hear Ernie open the envelope and sort of reads it and then he comes back and says to Dave I think there's a problem here somewhere Dave. Dave says what do you mean? And Ernie says the cheque seems to have fallen out! You can just imagine Dave Kemp, we couldn't see but we could imagine his jaw dropping."

Once all of the equipment and systems were configured, the station conducted an H-45 (Horizon minus 45 minutes) interface and testing with both Goddard Space Flight Centre and Mission Control Centre (MCC) at Houston. The quite complicated H-45 sequence is shown in Appendix 6.

Then before each pass in the group, the station carried out a shorter H-10 (Horizon minus 10 minutes) test and interface period with MCC. There was about 90 minutes between each pass in a group and for the first flight about 16 to 18 hours between each group. This gave the specialist team just enough time to go home, get some rest, and return in time for the three-hour setting up period for the next group of passes. A group of five passes took about seven and a half hours to complete then sometimes there was up to another hour of post-pass activity. While a flight was in orbit it was a tight schedule.

STS-1 launched from pad 39A of Kennedy Space Centre in Florida, USA, at 12:00 UTC on 12 April 1981, for a flight of two days and six hours covering 36 orbits.

Just prior to the first Orroral pass of the Space Shuttle, the crew had performed the Orbital Manoeuvring System (OMS) -2 burn of the rockets to place the Shuttle into the correct orbit. This first pass for Orroral occurred about one hour after launch and was considered to be a 'good' pass. That is, it was in a good position regarding the station horizon and it came up to a good elevation, about 45 degrees. However, the antenna predictions sent to the station were based on everything about the OMS-2 burn being correct. There was no time or means of obtaining an actual accurate orbit calculation before Orroral's pass, so the antenna operator was working with pointing information that may have had an unknown error component. There was also some doubt about what might happen because the previous tracking station, the Indian Ocean Station in the Seychelles Islands, did not acquire the S-Band signal. The Orroral shuttle tracking team were tense. However, when the spacecraft came over the horizon at Orroral, almost immediately, the receiver operators were able to announce confidently:

"Receivers have acquisition of DFI and OD!"

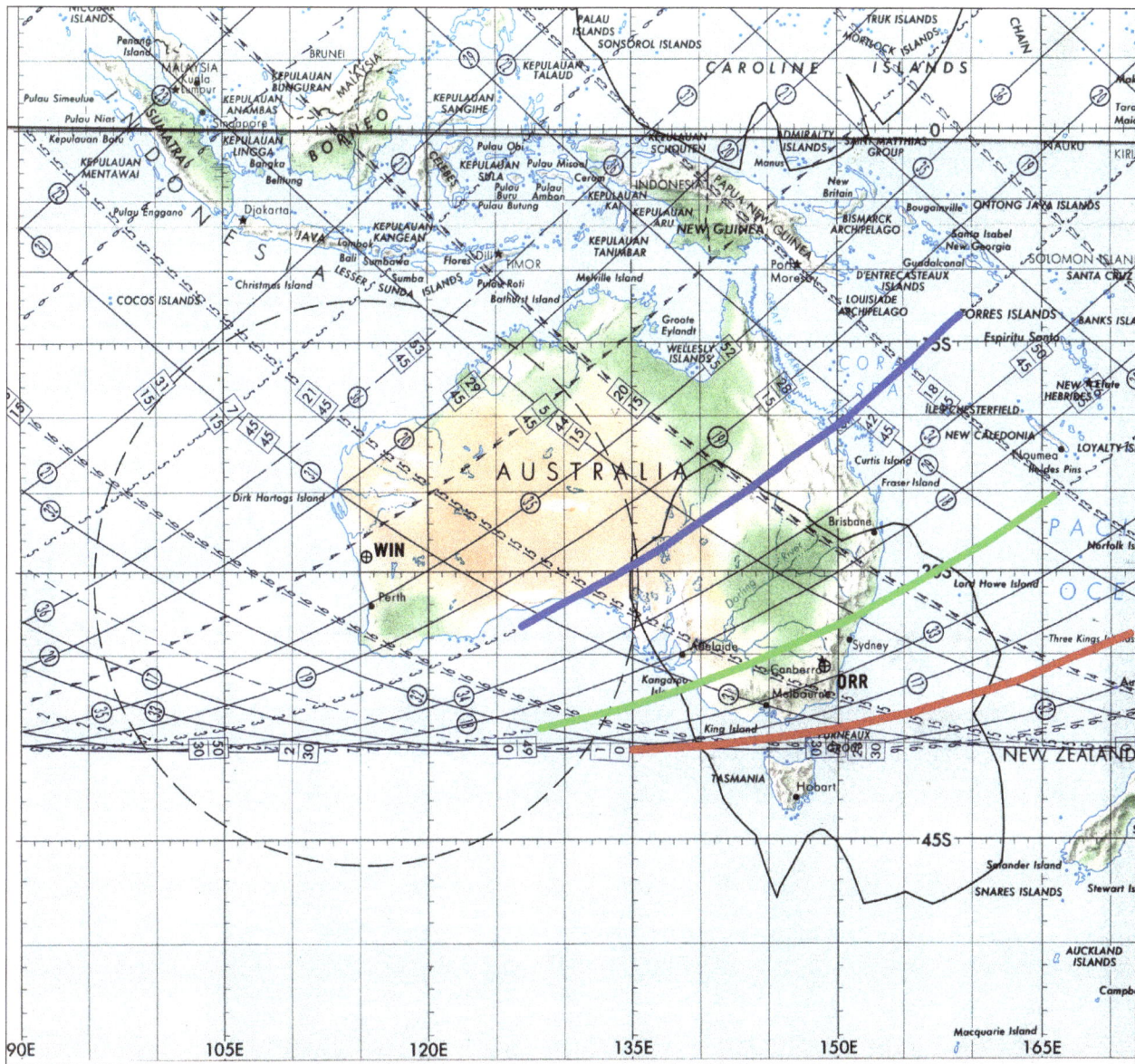

The first three orbits of STS-1 over Orroral Valley are shown above. Orbit 1 is red, orbit 2 is green and orbit 3 is blue.
The irregular black outline is the station horizon profile and shows orbit 3 as a very short pass.

Ken Strickland -

"I remember on the first pass when they didn't get it over the Indian Ocean Station and we did get it and the sense of relief and excitement at the time when we were able to lock up the receivers."

The staff relaxed a little and the shuttle was tracked without any further difficulty.

The pass was about 2 minutes and 45 seconds long. The spacecraft came over the south-western horizon, passed to the south of the station, and departed over the eastern horizon almost due east of the station.

There was no voice contact between Houston and the Shuttle while over Orroral during this pass.

Darryl Fallow -

"I think a high point for me at Orroral was the launch of the space shuttle. There had been a long period of simulation leading up to that. The noise of the shuttle going up is something I remember vividly. We had the voice conference up and were listening to the launch countdown. When the Shuttle was launched there was the sound of the thing, a bloody big roar. The thing took off and went over the first few stations. Then it went over the Indian Ocean station and they did not acquire the S-Band, so we were a bit jittery. But then for us everything just went like clockwork! We did get the S-Band and it was much easier than the simulations. The simulations were much harder by comparison. One other thing I recall is that given the speed that it was moving and the high data rates there was very little room for error, everything had to be spot on, which it was."

STS-1 lifts off from pad 39A at Kennedy Space Centre on 12 April 1981.

STS-1 landing at Edwards Airforce base 14 April 1981.

Here is a transcript from a recording of the voice communications at Orroral for orbit 16 of the first flight of the Space Shuttle. This part of the tape is a recording of the three communications circuits recorded on a single-track recorder. Therefore, there are a few occasions where conversations overlapped and could not be properly deciphered. The communications circuits recorded were:

1. The off-station communications between Orroral, the control centres in the United States, and other stations (External circuit).
2. The air to ground number one circuit between Orroral and the MCC Houston (Air/ground circuit).
3. The internal on-station communications circuit within the Orroral Valley Tracking Station.

As far as possible this transcript follows the actual conversations on these circuits in correct time sequence. Most call signs are self-explanatory, but where necessary definitions have been given.

During this pass, to honour Australia, the crew of Columbia played an excerpt from 'Waltzing Matilda' sung by Slim Dusty.

REC: is Orroral receiver technician

USB is Orroral unified S-Band technician,

DH is Orroral data handling technician,

Opsr is Orroral Operations Supervisor, Peter Uzzell.

Except where noted as GCC, the COMTECH is Philip Clark. (Author).

RTC is Houston Real Time Command controller,

DFE is Houston Data Flow Engineer.

NST is Goddard Network Support Team.

GCC is the communications centre operator at Orroral tracking station.

QUITO is the call sign of Quito tracking station in Ecuador.

(The call sign 'M&O' was used by the on-station operations controller to commemorate the call sign of the station operations controllers that were originally used only at the earlier manned spaceflight tracking stations.)

Tape Time (Min:Sec)	Callsign	Transcript

Communications on external circuit

21:02	GO	Quito Goddard ops site co-ord.
21:05	Q	Quito
21:09	GO	OK Quito, I'm going to release you at this time, like to thank you for your support.
21:16	Q	Roger

Communications on air/ground circuits

21:16	HOUSTON COMTECH	OK you're load & clear, stand by.
21:20	Y	Roger.
21:25	HOUSTON COMTECH	(With tones) This is Houston comtech test one, two, three, two, one, end of test.
21:35	Y	Houston comtech Yarragadee comtech we have 100% keying modulation go.

Tape Time (Min:Sec)	Callsign	Transcript
21:39	HOUSTON COMTECH	Roger Yarragadee.
21:40	ORRORAL COMTECH	Orroral comtech we have 100% keying air/ground one air/ground two.
21:45	HOUSTON COMTECH	Roger Orroral, thank you. Configure for your pass.

Yarragadee talking on external circuit.

22:26	Y	Yarragadee is configured for real-time support,
22:28	HOUSTON COMTECH	Roger.

Communications on external circuit

22:40	RTC	Orroral RTC.
22:42	Opsr	We are configured for command interface.
22:36	RTC	RTC copies.

Yarragadee talking on external circuit.

22:56	Y	Yarragadee is go for voice.

SPACE SHUTTLE is through Yarragadee.

23:18	CAP COM	Good morning Columbia this is the Crimson Team through Yarragadee, we'll be with you for 8 minutes. How do you read? Over.
23:53	CAP COM	Good morning Columbia this is the Crimson Team through Yarragadee, how do you read over.
24:00	SPACE SHUTTLE	Good morning Crimson Team we read you loud and clear. How you doing Joe
24:03	CAP COM	OK good morning you are very very weak. We've got nothing special for you. Except to say we are happy with the PCS config as it is right now.
24:17	SPACE SHUTTLE	Good morning Joe how you guys doing it's about time you all came to work.
24:22	CAP COM	We've just been watching and enjoying. We are proud of the Silver Team though. They did a grand job and so did you. We are thinking of having them bronzed, in fact.
24:46	SPACE SHUTTLE	(Garble)
24:49	CAP COM	OK, John and Crip, you are very weak, and we may have some comm problems if we don't get much to you this pass, we'll be back very shortly through Orroral Valley.

Yarragadee talking on external circuit.

25:08	HOUSTON COMTECH	Yarragadee comtech Houston comtech.

Tape Time (Min:Sec)	Callsign	Transcript
25:11	Y	Houston comtech Yarragadee comtech.
25:14	HOUSTON COMTECH	Are you having transmission problems?
25:15	Y	Negative. No I have no indication of problems, it is just weak signal.
22:24	Y	Would you like me to remove the squelch?
25:29	HOUSTON COMTECH	Negative Yarragadee, stand by.
25:32	Y	Roger.

SPACE SHUTTLE is through Yarragadee.

25:14	CAP COM	Columbia it is not comm problems, you are just weak in your transmissions to us.
25:48	SPACE SHUTTLE	(faint) it is ... looking at this pressure . called the 02. Oh, you were happy with it. I missed that part.
26:04	CAP COM	OK. Crip. We are happy with the current PCS configuration. We are looking at that reg pressure and we will keep you advised on that.
26:15	SPACE SHUTTLE	Ok. You sure are easy to please.
26:21	CAP COM	Well, we may not be when we see some data here in a few minutes. But we are keeping a careful eye on it. There is nothing that can break as we watch it so we are not particularly worried.
26:35	SPACE SHUTTLE	It seems to have levelled off at around 215 or so.
26:40	CAP COM	OK. We copy that. Thank you.
28:12	SPACE SHUTTLE	I finally got around to my first cup of coffee. Sure tastes good
28:22	CAP COM	Roger that.
28:25	SPACE SHUTTLE	Not really a cup though.
28:45	SPACE SHUTTLE	... said something else.
28:49	CAP COM	Crip you are dropping out here.

Orroral & Yarragadee talking on external circuit.

29:20	DFE	Orroral DFE.
29:21	Opsr	Orroral.
29:22	DFE	Your lines are enabled would you confirm you are green for support?
29:26	Opsr	That's affirmative, do we have a clearance for destination code 150?
29:30	DFE	That's affirmative on that Orroral.

Tape Time (Min:Sec)	Callsign	Transcript
29:46	Opsr	Orroral destination code 150.
29:49	DFE	Copy that, we do see the 150 thank you.
30:36	Y	Yarragadee has one minute to LOS.

SPACE SHUTTLE is through Yarragadee.

30:46	CAP COM	Columbia, Houston. We are 30 secs. from LOS. We will be gone for a minute and a half and be back with you at 23 + 22.

Yarragadee talking on external circuit.

31:09	Y	Houston comtech Yarragadee comtech we copied no downlink.
31:13	HOUSTON COMTECH	Roger.
31:37	Y	Yarragadee has LOS.
31:41	HOUSTON COMTECH	Roger Yarragadee.

Communications on external circuit

33:06	REC	Orroral has acquisition OD 22:17.
33:10	DFE	DFE copies.
33:12	REC	Orroral has acquisition DFI 22:23.
33:20	USB	Orroral is go for command.
33:26	USB	Ranging initiated Orroral.
33:42	REC	Orroral has acquisition FM carrier 22:51.
33:46	DFE	DFE copies.
33:49	Opsr	Orroral range acq is complete, range delta is 0.9 microseconds.
33:57	REC	FM dump 22 er 23:07.
34:02	DH	PCM lock on dump FM, 960 reverse.

SPACE SHUTTLE is through Orroral

33:35	CAP COM	Hello, Columbia. This is Houston back with you through Orroral Valley. We will be with you for 3.5 min and can report that the IMU cal has been completed.
33:46	SPACE SHUTTLE	OK. Fine and dandy. You got any other traffic for us?
33:50	CAP COM	Not much, Crip. You are loud and clear on this pass. Curious to know if you have message 11 aboard. It is a pretty major change to timeline and prepare to answer questions, when and if.

Tape Time (Min:Sec)	Callsign	Transcript
34:05	SPACE SHUTTLE	OK. I'll tell you what. John is down on the mid deck now. We will check that out for us. Meanwhile I got a little Slim Dusty and Waltzing Matilda for our friends down under here.
34:16	CAP COM	Let her rip.
34:22	SPACE SHUTTLE	**(Music - 'Waltzing Matilda' sung by Slim Dusty– 37 seconds.)**
34:59	SPACE SHUTTLE	Too bad it is always dark when we are going over. We can't get a good view of it.
36:35	CAP COM	Aw, but they got it, a good sound of it. I think the S-band will never be the same again.
35:12	SPACE SHUTTLE	Probably not. probably not.

Communications on external circuit

36:01	ORRORAL COMTECH	Houston comtech Orroral comtech site co-ord.
36:04	HOUSTON COMTECH	Houston comtech.
36:12	ORRORAL COMTECH	Roger, we copied a short burst of 450 HZ on air/ground 2. It's gone again now.
36:12	HOUSTON COMTECH	Roger.
36:23	HOUSTON COMTECH	Orroral comtech Houston comtech site co-ord
36:25	ORRORAL COMTECH	Orroral comtech.
36:27	HOUSTON COMTECH	You did copy Waltzing Matilda did you not on the air to ground?
36:31	ORRORAL COMTECH	That's Affirmative, we almost recognised it.
		(Explanatory Note: The complex sounds of the music were very badly distorted by the severe limitations of the air to ground voice link between the spacecraft and the ground. This link was specifically designed for voice communications only.)
36:40	HOUSTON COMTECH	Ha, Ha, roger that.
36:53	REC	FM dump off 25:58.
36:57	DFE	Thankyou.
36:58	Opsr	Approximately 1 minute to horizon LOS.
37:14	REC	LOS FM 26:17.

SPACE SHUTTLE is through Orroral

37:17	CAP COM	Columbia, we are about 30 seconds from LOS. We'll be back with you at 23 plus 54.
37:26	SPACE SHUTTLE	Roger, 23:54.

Tape Time (Min:Sec)	Callsign	Transcript
37:29	CAP COM	And we enjoyed the music, Bob, Thank you.
37:32	SPACE SHUTTLE	Oh, we enjoyed it. We just wanted to share some with you.

Communications on external circuit

37:51	REC	LOS DFI 26:55.
37:56	DFE	DFE copies.
38:02	REC	LOS OD 27:06.
38:07	DFE	Copy that Orroral.
38:24	Opsr	GC Orroral.
38:27	GC	Orroral GC.
38:29	Opsr	We'd like to convey our thanks to the Columbia crew for the little bit of hometown music.
38:35	GC	Well, OK We'll pass that on.

END OF ORRORAL PASS RECORDING

(GC is the Houston operations controller.)

Richard (Dick) Elliott recalled a pass on a later flight -

"I was in the data handling area. There was this incident with another technician when my attention was distracted. He had a cup of coffee and had idly made some holes with a pencil about half an inch below the rim in the polystyrene cup. He was sort of doodling. Anyway, his job was to operate the data handling for another scientific spacecraft while I was doing the Shuttle. Then just when I was about to do the Shuttle pass he picked the cup up by the wrong side and most of the coffee piddled out the hole he had made and piddled down the front of his shirt. Seeing this done almost in slow motion as he was trying to drink it and he realised he was drowning as well made me have the giggles that just about disabled me for about an hour and a half afterwards. Fortunately, nothing went wrong, but it was very difficult to concentrate after having seen such a stupid thing been done. Every time I looked at him afterward still with the mess down his front, I could not help but get the giggles! I remember another time one of the 642B computers failed, but it was just a matter of switching the big 1299 switches over and having the backup computer come up. It was really just a breeze because we had done all this in simulations."

The first four flights of the shuttle used the orbiter 'Columbia' and they were of durations ranging from two to eight days. Only a crew of two was carried on these. They were classed as research and development flights. A single specialist team was used at Orroral for tracking these flights.

Facilities supervisor **Trevor Smith** recalled an incident during one of the shuttle passes:

"On one of the first passes that got up there we were actually tracking on the 9 M antenna. It had a heat exchanger, and there was a tracking pass due to come up in a very short period of time. The main contactor that controlled the fan on the heat exchanger failed just before the pass. The antenna could not be operated without the heat exchanger functioning. The coil had failed, so there was a mad panic. I said to Dave Kemp, while this track is going I will go up and physically hold the contactor in. He said you can't do that, but I said 'yes I'm going to or we will miss the pass'. So I went up the antenna and physically held the contactor in for the track. We were able to do the pass and those passes were 90 only minutes apart, and in that 90 minutes I changed the contactor over and then we were back to normal tracking."

After the successful first flight of STS-1 the astronauts John Young and Bob Crippen visited the station in late 1981 and made presentations to staff commemorating the flight.

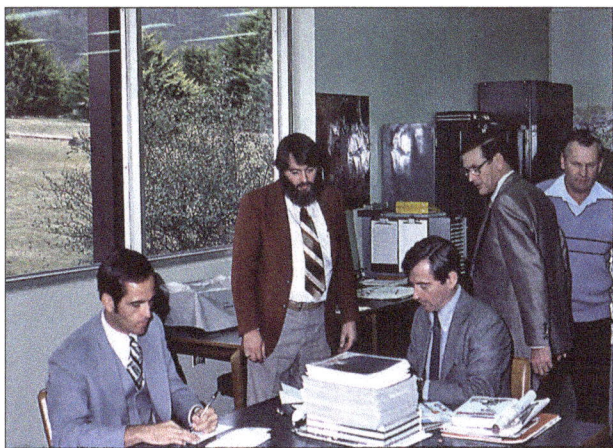

Signing mementoes for station staff. L - R: Astronaut John Young, Author Philip Clark, Astronaut Bob Crippen, Engineer Operations Analysis Richard (Dick) Simons, Shuttle Team Leader and SOS Peter Uzzell.

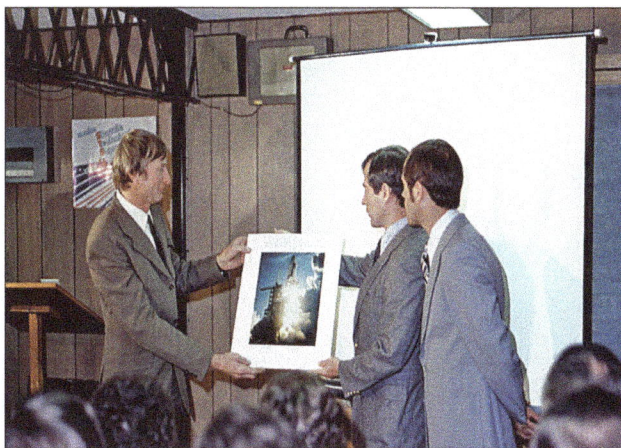

Bob Crippen makes a presentation to Station Director, Lewis Wainwright.

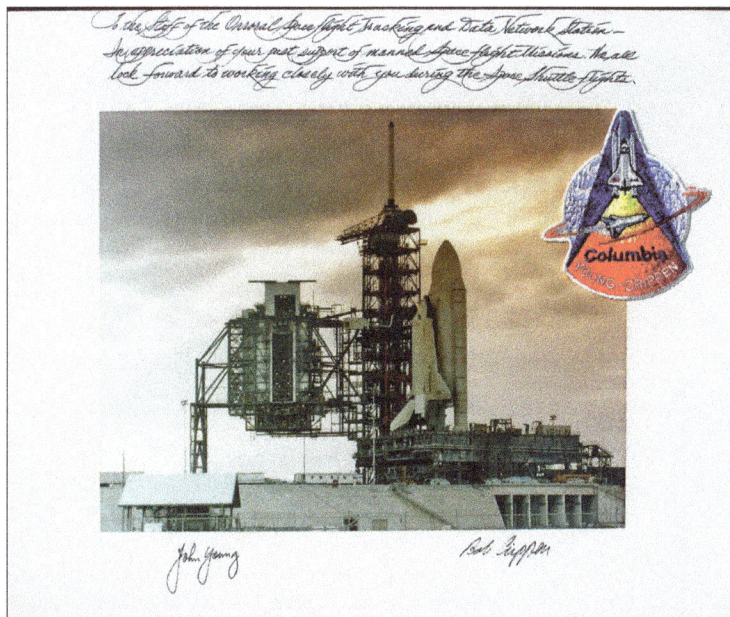

This framed and signed award was presented to the station by astronauts John Young and Bob Crippen. Only two copies are known to exist.

The cooking staff in the canteen provided a sumptuous lunch with a very striking cake depicting the Space Shuttle.

The Shuttle cake

Preparing lunch for the astronauts and guests.

Lindsay Richmond remembers John Young coming to the station with a piece of thermal tile from the shuttle. The piece of tile was only about four centimetres cube. He saw them give a demonstration when one of the facilities technicians heated the tile with an oxy-acetylene torch until the one end was red hot and it was still safe to touch the other end with bare skin. I (author) still have that piece of tile.

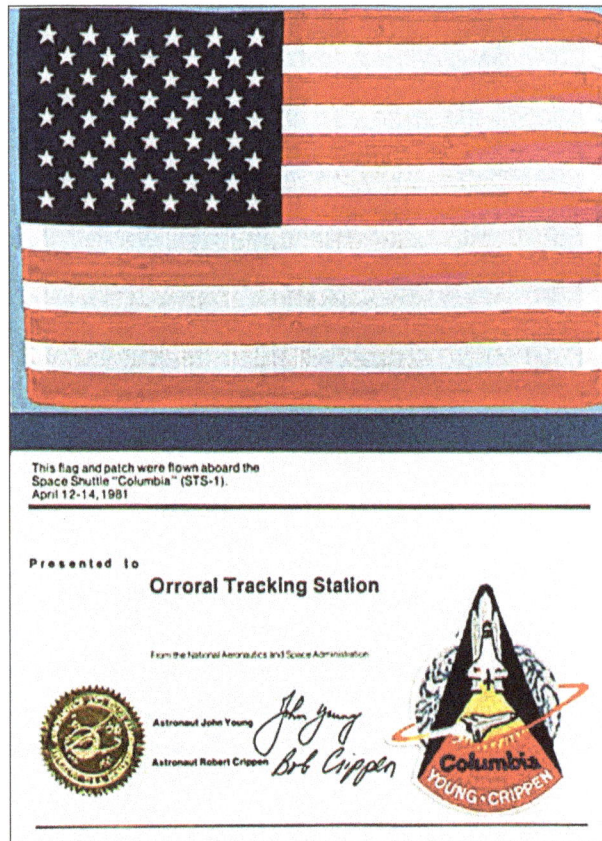

This flag and patch were flown aboard the Space Shuttle "Columbia" on the STS-1 flight 12 - 14 April 1981.

Patches for all of the Space Shuttle flights tracked by Orroral Valley. Each patch supposedly has a symbolic representation of the flight number as part of the design. Can you find them?

For a few of those later flights, where much of the support period at Orroral spanned evening and night shifts, and the groups of passes were unfavourable, two specialist operations teams were assigned. This was to reduce fatigue and dangerous driving conditions for the teams going home for rest. Such was the case for Orroral support of STS-8 and STS-9 flights.

Dave Richards -

"There was the time that we stopped a space shuttle mission. We stopped the countdown worldwide. They were going through the countdown, fortunately not too far into it, and suddenly the joystick on the 85-foot antenna would not work. We informed NASA and they stopped the countdown while we went to the store to have a look for a spare part. Meanwhile NASA were asking every few minutes an estimated time when our equipment would be working again. I think it was Lindsay Richmond was helping me to look in the stores and he came back and said we haven't got one. He said it's been thrown out because there was this rationalisation of stores stocks. As he came out of the stores he happened to look behind the door and saw this cardboard box. It turned out that in this box was one of the special potentiometers we needed. I was able to whack this in, it took me about 10 minutes, and then they were able to start the countdown again. As Orroral was critical to this particular launch, if we had not found a part and they'd had to get one sent out, it would have taken so long that they would have had to empty the tanks on the space shuttle. That would have been a hugely expensive operation. So, I've got the notoriety of being a person that has stopped the countdown worldwide."

STS-9 was a significant flight for Orroral for two reasons. First, it was the longest flight that Orroral supported. Second, it was the flight for the only known official amateur radio experiment. This is documented in Chapter 9. Because it was a longer flight, 10 days and seven hours, and also spanned night and evening shifts at Orroral, two specialist teams were used to supplement the normal shifts for the tracking of this flight. During this flight Orroral supported 29 orbits.

Until it ceased tracking at the end of 1984, Orroral tracked all of the Space Shuttle launches. The final track of a manned spacecraft at Orroral was STS-51A, the 14th Space Shuttle flight, in November 1984.

Chapter Nine - The Amateur Radio Experiment

When the amateur radio technicians at Orroral Valley Space Tracking Station learned that payload specialist Dr. Owen Garriott W5LFL, was to take a hand-held amateur radio with him on the STS-9 flight, they proposed that an official amateur radio experiment with the Space Shuttle be conducted. The technicians were aware the NASA placed a very high priority on being able to speak directly with the astronauts on the Space Shuttle as much as possible. The experiment proposed was to show that by only using amateur radio and no NASA space tracking systems, in the event of an emergency the astronauts could be linked back to MCC at Houston. Most of the planning and coordination for this experiment was done by Richard (Dick) Elliott (VK1ZAH).

AMATEUR RADIO TO FLY ON STS-9

Mission Specialist Owen Garriott will use a hand-held radio during part of his off duty time in the STS-9 mission to communicate with some of the thousands of Amateur Radio operators around the world.

While only licensed Amateur Radio operators will be allowed to transmit signals, anyone with one of the popular programmable "scanner" radios can listen to these conversations.

Dr. Garriott's transceiver is a battery powered unit capable of five watts of output. The printed circuit antenna is placed on *Columbia's* upper aft flight deck crew compartment overhead window. Garriott will wear the standard in-flight headset when operating the radio.

All radio operation for the STS-9 will be in the Amateur Radio 2-meter band, 144 to 148 MHz. The primary frequency Dr. Garriott will be using when transmitting over the U.S. is 145.55 MHz. All that will be needed to listen is a receiver (such as a scanner) capable of tuning to 145.55 MHz. The backup frequencies are 145.53 and 145.57 MHz.

When *Columbia* approaches the portion of the ground track where Amateur Radio operations are planned, Garriott will call and transmit continuously for one minute on the even minutes and will receive continuously for one minute on the odd minutes.

During a typical even minute transmission period, Dr. Garriott will identify a geographical area that he will listen for. He will also, as time permits, describe the flight crew activity or views of Earth.

During the odd-minute receive period, Garriott will scan the announced receive frequencies for call signs from the designated area only. To establish contact, an Amateur Radio operator will send his full call sign only, repeating it several times during the scanning period.

During the next transmission, Garriott will acknowledge all call signs he has heard during the listening period. No other report will be needed: call-sign identification constitutes a two-way contact. This procedure will give more operators a chance to make contact. If time permits, some stations may be called on for short transmissions to fill the time period.

Dr. Garriott's call sign is W5LFL. Use of the transceiver will be limited to one hour a day.

Part of the NASA pre-launch press release for STS-9

The first problem to be overcome was that of actually setting up the experiment itself. At the time the NASA representative in Australia was astronaut Dr. Joe Kerwin, who fortunately was based in Canberra. When he was approached by the amateur radio technicians from Orroral, he was very receptive to the idea. He was in contact with his colleague, Owen Garriott, and he suggested the basis of the experiment to him.

Dr Kerwin then came back to the Orroral amateurs and told them that Dr Garriott was very keen on the idea. He asked them if they were up to the challenge. They readily accepted, and with the help of the astronauts the plan was accepted by MCC at Houston. In November 1983, during the planning, Dr Garriott expressed to Joe Kerwin how much he wanted to contact the amateur radio people at Orroral during his flight. This gave the experiment an added boost.

However, there was a further problem. The original plan for the experiment was that it would be conducted from an amateur radio station established at the Orroral Valley Space Tracking Station. The overall management of the tracking station was under the Station Director, the representative of the Department of Science and Technology. The Station Director, after consultation with his department, was concerned that the amateur radio equipment would interfere with the station's normal satellite tracking operations.

Astronaut Dr Joe Kerwin at Deakin

Although the amateur radio technicians had carried out unofficial tests to prove their equipment did not interfere with normal tracking operations, the Department was firm in that it would not permit the amateur radio station for this experiment to be established at Orroral Valley. Thus, another site had to be found for the experiment ground station. After considerable negotiation it was agreed by all parties that the suitable site would be the NASCOM voice and communications switching centre in the Canberra suburb of Deakin. Before the launch of the STS-9 flight, the station was set up at Deakin, with the antennas on the roof of the building.

Hauling antenna components onto the roof of Deakin Switching centre.

The antennas used for the STS-9 amateur radio communications.

To this point the ground-work and liaison with Dr. Kerwin had been done by Richard (Dick) Elliott, VK1ZAH. He spent considerable effort in planning the amateur radio experiment, and working out which suitably licenced staff would be available to operate the amateur equipment. Following are some of the documents used in the planning and working out people available to operate the station. Because of the number of amateurs involved, the special nature of the station, and the participation of NASA through Dr. Kerwin, it was decided that a special call sign for the station should be obtained if possible. The Department of Communications was contacted, and thanks to their understanding and ready co-operation when the experiment was explained, they issued the special event call sign of VK1ORR for the duration of the STS-9 mission.

Subject: Amateur Operation with STS9.   15/11/83

A new twist has occurred because Joe Kerwin has received a reply from Garriot saying how much he wants to work Orroral staff & possibly Joe Kerwin. What callsign should he call etc. The department in its wisdom still will not permit operation from Orroral but wish the Orroral hams to provide a ~~transm~~ transceiver site at Deakin Switch.

It is intended that a limited or full call will make the contact then allow Joe Kerwin to talk to the astronauts, and let the other Orroral hams have a go too.

To make things easier we intend to be told in advance what orbits are likely and what frequency (2 meters FM simplex) to use. This puts us at a great advantage over other hams, and makes a QSL card at least a good possibility.

Because of the 2 ~~team~~ Shuttle teams having ~~two~~ hams on them we will need to work out a roster of people who can staff Deakin. I would like the novice call holders who can attend to be there as well.

I propose to forward my callsign VK1ZAH by telex as the one to be used at Deakin. Any contact and I guess there could be more than one orbit ~~at~~ will probably occur late in the mission (2nd half anyway) although some clever person is now describing the Ham link as backup voice, which aids our cause.

Looking at the proposed shuttle roster & staff disposition

A shift ZJF, ZAE, NDM, 2 ????   outside 0100—1400 either 3/11/83
D shift DF   outside 0400—1202 until 3/11/83
F Troop ZQR, 2NKP outside 2000—????Z
S Team 2 ZAH, RR   outside 1400—2000Z
S Team 1 2KPG   outside 0400—1400Z

Assuming operators were picked up / put down at Deakin an hour before or after shift!

If you can interpret the attached page it appears that considering the later ½ of the mission Deakin can be manned by F troop almost always, D shift most of the time, A shift & ST1 on the 'midnight' shift passes and ST2 on the afternoon shift passes.

Please advise me — Dick Elliott of your proposed interest & participation.
                                        IMMEDIATELY!

Launch on MET (a.m.) 1-8-83
GMT Based on L/O 001128/1630Z.

Local time to Eastern daylight saving time.

ORBIT	MAX.EL.	GMT AOS	LOS	GMT H-45	GMT SRT/RELEASE	LOCAL.
3	6.2	28/20391	20415	28/1954	28/1654	29/0354
4	46.1	29/			28/2325	29/1025
14	14.0	29/1241	12	29/1146	29/0856	29/1956
15	13.8	29/1414			29/1528	29/0228
19	8.8	29/03021		29/1945	29/1645	30/0345
20	15.6	29/			29/2316	30/1016
30	21.8	30/1232		30/1127	30/0847	30/1947
31	7.7	30/1507			30/1517	01/0217
35	12.4	30/0320	20	30/1936	30/1636	01/0336
36	15.7	30/1532			30/2307	01/1007
46	37.4	01/22042		01/1135	01/0835	01/1935
					01/1546	02/0046
51	17.4	01/ 1010		01/1925	01/1625	02/0325
52	10.0	01/1455	21		01/2256	02/0956
62	66.4	02/1210	12	02/1125	02/0825	02/1925
					02/1327	03/0027
67	25.7	02/20002	20	02/1915	02/1615	03/0315
68	6.1	03/21335	21		02/2246	03/0946
78	61.7	03/11593	12006	03/1114	03/0814	03/1914
					03/1316	04/0016
83	41.3	03/19497	19555	03/1904	03/1604	04/0304
					03/2104	04/0704
94	35.1	04/11485	11551	04/1104	04/0804	04/1904
					04/1305	05/0005
98	3.1	04/13070		04/1722	04/1422	05/0122
99	72.4	04/193817			04/2053	05/0753
110	23.8	05/113726		05/1052	05/0752	05/1852
					05/1254	05/2354
114	4.6	05/175511		05/1710	05/1410	06/0110
115	59.0	05/19267			05/2041	06/0741
125	10.2	06/09552		06/0910	06/0610	06/1710
126	16.4	06/11262			06/1141	06/2241
130	6.4	06/17422		06/1627	06/1357	07/0057
131	36.6	06/19145			06/2028	07/0728
141	14.5	07/09425		07/0857	07/0557	07/1657
142	12.2	07/1115			07/1228	07/2328
PLANNED EOM						
146		08/1730		08/1615	08/1345	09/0045
147		08/1905			08/2016	09/0716

With the preliminary arrangements completed, the site established and the call sign allocated, the equipment and antenna systems were selected. The choice of antenna was not simple because of the conflicting requirements that had to be met. By this time launch day was approaching and not a great deal of time was available to construct special equipment. As this Shuttle flight had a high inclination orbit (57 degrees to the equator) an omnidirectional antenna would have been desirable. Such an antenna should have a low angle of radiation to cover the low-elevation horizon portions of the pass, and a higher angle for the maximum elevation point. These characteristics are, for all practical purposes, mutually exclusive.

In addition, physical size and weight had to be kept within manageable limits as the entire assembly had to be man-handled up a ladder onto a roof some 15 metres above ground. The speed of a fast-moving spacecraft posed problems for a directional antenna system. Finally, after consideration of all factors, a combination was selected to cover as many possibilities as was reasonable. This combination was a steerable crossed 10 element yagi with switchable circular polarisation, a 5-element horizontal yagi oriented to maximum elevation of the pass, and a 5/8 wavelength vertical whip. The antennas were mounted on a temporary scaffold erected on the roof of the Deakin Switching Centre.

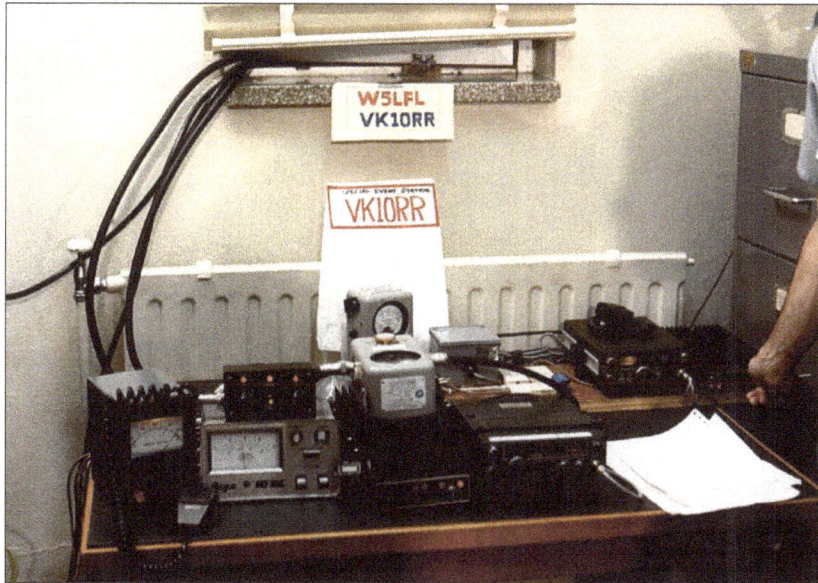

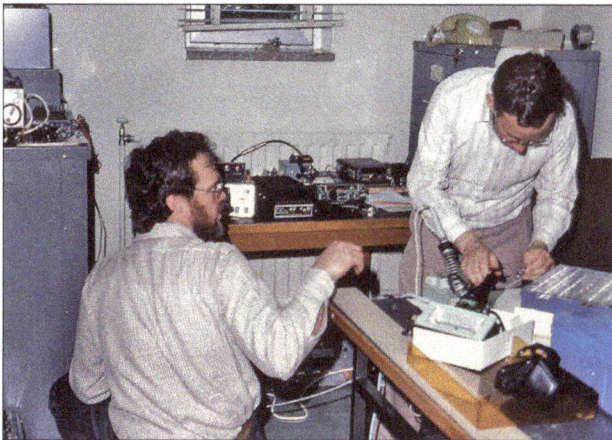

Richard Elliott (left) and Darryl Fallow set up equipment.

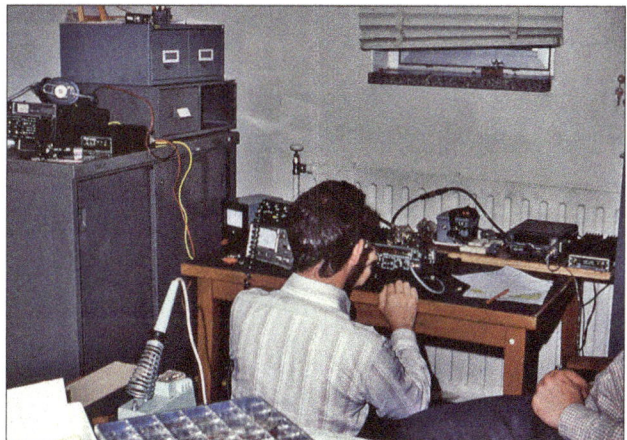

Darryl Fallow makes some adjustments.

Semi-rigid, low-loss coaxial hard-line was used to connect the antennas to the equipment. A low-noise GaAs FET pre-amplifier was used to improve the receiver noise figure. All of the equipment used was provided by local amateurs. It was configured in two chains to give redundancy in the event of a failure. Alternative mains and battery power was available to all essential equipment. Three transceivers were used in the final configuration. These were an ICOM IC260A, a FDK 750A and an ICOM IC251A. The two main chains used receiver pre-amplifiers and had Microwave Modules 100-watt linear amplifiers for transmitting. This configuration allowed two 100-watt uplink paths.

The prime receive path threshold was -140 dbm, due to the GaAs FET amplifier at the antenna head. Special delayed transmitter keying was installed to enable the antenna-head GaAs amplifier to be disconnected before uplink power was applied to the antenna. Thanks must go to Richard Elliott VK1ZAH, Paul Bell VK1BX, Darryl Fallow VK1DF, Bob Henson VK1RR, and Bob Quick VK1ZQR, for their efforts in construction and installation of the equipment. The officers in charge of the Deakin Switching Centre, Mr. Des Terrill and Mr. John Warth, provided valuable assistance in making space available and advice during installation of the station. For the actual contact, and to prove that the amateur system could provide proper back-up voice capability to the Space Shuttle, a telephone patch was used to interface the radios with the normal telephone system. Dr. Owen Garriott W5LFL, on board "COLUMBIA" was then able to speak to the CapCom (Spacecraft voice communicator) in Houston via amateur radio through VK1ORR and a phone-patch interface unit. On Monday evening 5th December 1983, this historic test took place during orbit 110 of the STS-9 mission.

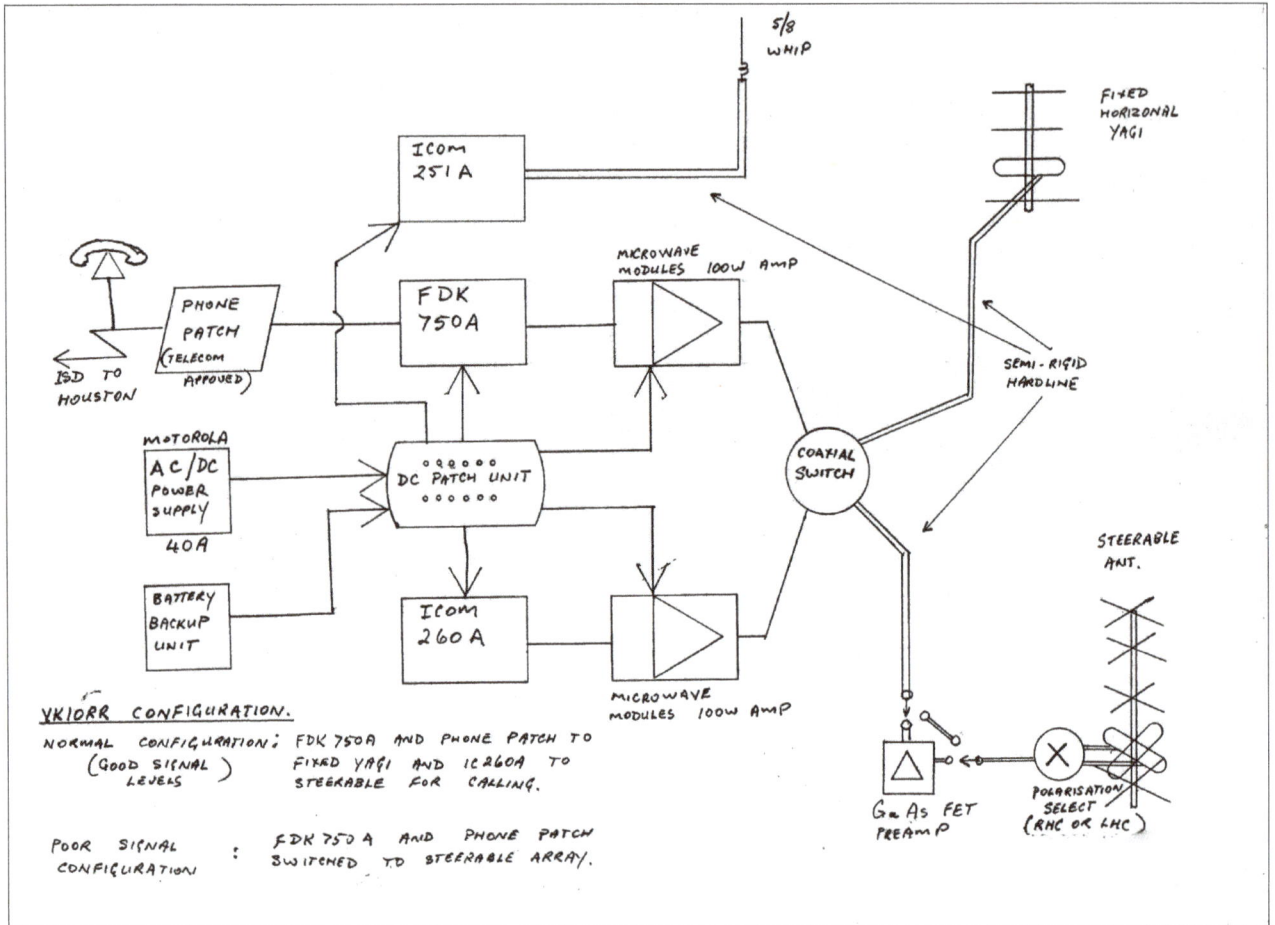

The configuration of VK1ORR

The test proved an outstanding success and demonstrated that amateur radio could provide excellent emergency voice communication from space. The orbiter was travelling from north-west to south-east over Australia and passed directly over Melbourne. This pass allowed only six minutes for the contact. During his conversation with controllers in Houston, Owen Garriott said of VK1ORR: "This is one of the best stations we have heard since we have been in orbit." A compliment indeed and a tribute to the performance of the VK1ORR station.

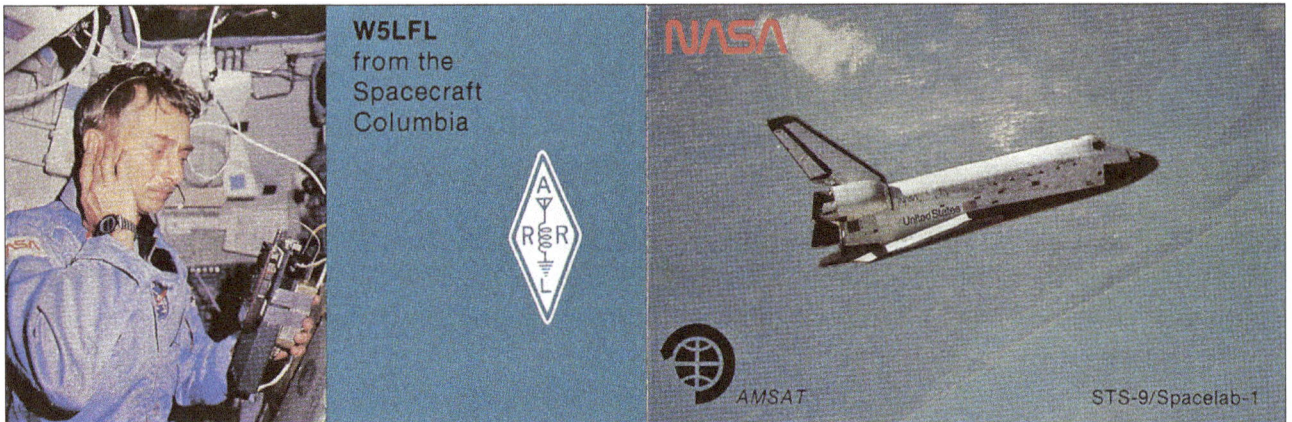

QSL card front.

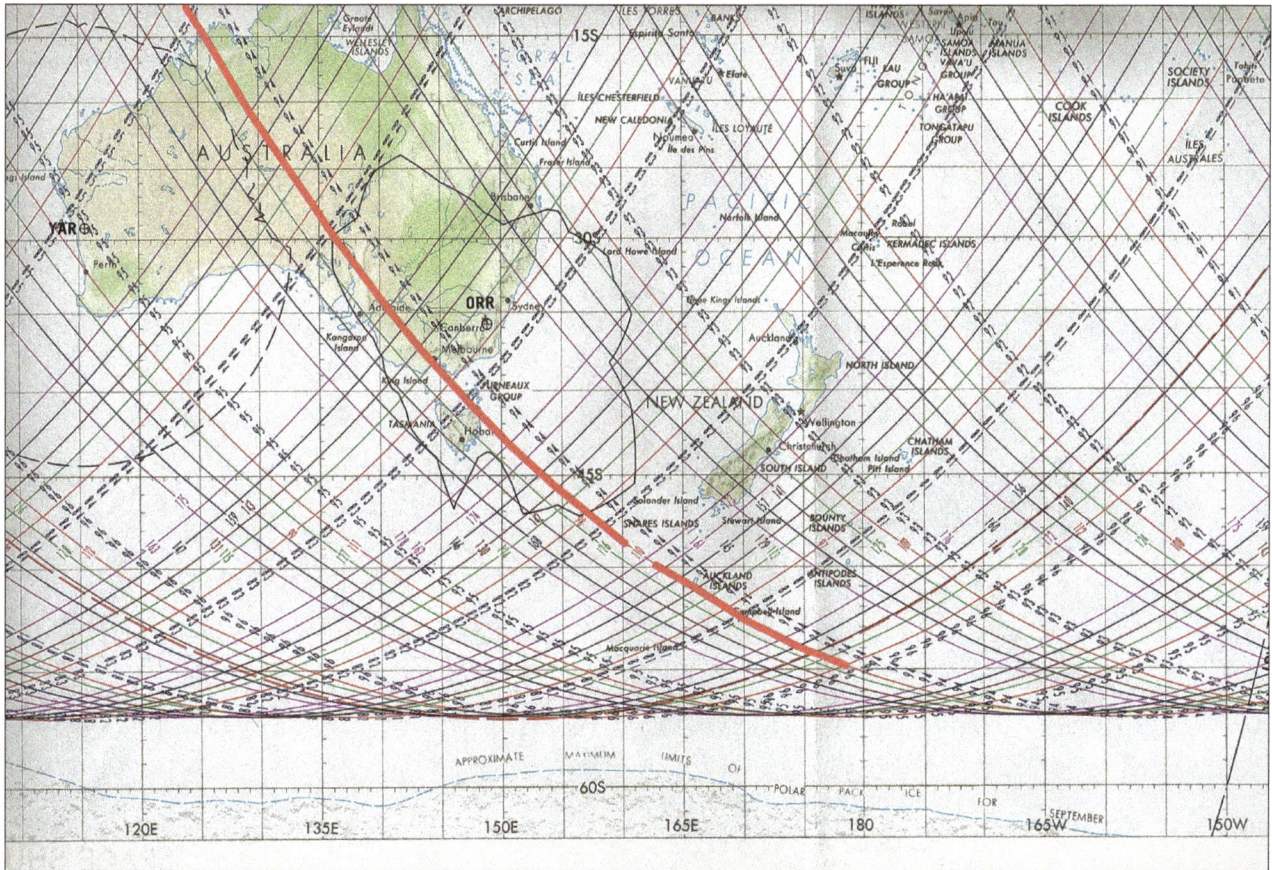

The ground track map shows orbit 110 in red as Columbia was passing over Australia from northwest to southeast.

Also present for this history-making experiment were the U.S. Ambassador to Australia, Robert Nesen, and Senator Jake Garn of Utah, U.S.A. a member of the NASA Appropriations Committee. The ambassador was able to exchange a few words with W5LFL during the contact which was coordinated by Dr. Garriott's colleague, Dr. Joe Kerwin.

FLIGHT OF COLUMBIA
STS-9/Spacelab-1

Launched on November 28, 1983
   and after 247 hrs, 47 min
landed at Edwards A.F.B. on December 8, 1983

- First launch of Spacelab (provided by the European Space Agency)
- Longest Orbiter flight to date
- First European crewmember
- First 'Payload Specialists' (non-career astronauts)
- First six-person spaceflight
★ First Amateur Radio station in space:
   W5LFL

Transceiver: modified Motorola MX-300 2-meter FM transceiver, hand-built by the Motorola Amateur Radio Club in Florida.
Antenna: directional ring radiator with cavity, designed to fit in the upper window of the spacecraft; built for NASA by volunteer employees of Lockheed.
Power: 4.5 watts
Mode: FM, CW (by keying carrier) All transmit and receive audio were tape recorded, which constitutes the station log.
Operating orbits: 40D, 56D, 62A, 71D, 91A, 96A, 97A&D, 110D, 111A&D, 112A, 113A, 129A, 130A, 134A, 134D, 135A&D, 144A&D, 145A&D, 146A, 149D and 150D.
Stations, 2-way contact: over 350
SWL: approximately 10,000 cards received
Countries: 23
Total operating time: about 4 hrs, 30 mins.

W5LFL/VK2KPG

Space Shuttle Columbia

This confirms a two-way contact on 2 meters with Scientist/Astronaut Owen K. Garriott, W5LFL, operating in space from aboard Columbia on the flight of STS-9/Spacelab-1 between 28 Nov. and 8 Dec., 1983.

73, Owen

Owen K. Garriott, W5LFL

Rear side of the author's QSL (radio contact confirmation) card for the STS-9 flight showing his callsign at the time, VK2KPG

Here is a transcript of the historic two metre band amateur communications test between W5LFL on board the Space Shuttle "Columbia", and VK1ORR in Canberra.

S/C is the Spacecraft "COLUMBIA", Dr Owen Garriott W5LFL.
ORR is VK1ORR in Canberra and
JK is the NASA representative, Joe Kerwin.

ORR	W Five Lima Fox Lima here is Victor Kilo One Oscar Romeo Romeo, over.
S/C	Hello Victor Kilo One Oscar Roger Roger, Victor Kilo One Oscar Roger Roger; W Five Lima Foxtrot Lima, W Five Lima Foxtrot Lima, I have you fives.
ORR	W Five Lima Fox Lima here is V K One Oscar Romeo Romeo. We copy you five by five. Operators here are V K 1 Zulu Alpha Hotel, V K 1 Bravo X-ray, V K 1 Delta Fox, V K 1 Romeo Romeo, V K 1 Zulu Quebec Romeo, and monitoring the s-band downlink V K 1 Zulu Alpha Hotel, V K 1 Zulu Item Foxtrot, V K 2 Papa Mike November, V K 2 Kilo Papa Gulf, V K 1 ZAE also. I will now hand you over to Joe Kerwin.
S/C	Columbia coming right back (signal fade) is Joe Kerwin at the station this morning? Over.
ORR	Roger Owen, we're copying you OK. We have had a problem with the phone patch. We're trying to get the phone patch up and then I'll hand you over to Joe.
S/C	OK, that sounds fine and I've got all your signals here on the tape recording and when we get back on the ground we'll be able to acknowledge them all in proper order and er, hullo there Joe, if you can get on the mike all your friends here in the spacecraft wish you hullo as do your friends back in Houston, and we will of course be looking forward to seeing you back home in a few more days and we also want to send our best wishes and many thanks to the fine crew there at Orroral Valley for all their help in tracking for the last week or so and the other previous missions, and an outstanding job has been done and we appreciate it very much. Go ahead now Joe, over.
JK	Alright, sounds like a real science party Owen, it sounds like you're getting a good time up there. I'm ready to turn - Have we got Houston on the line here? - OK Owen, I'm going to turn you over to Houston to complete the contact right now and then I want to get back with you for a minute. Please finish your transmissions with 'over' as we are simplex and I have to push the button for Houston to talk. Go ahead and call Houston, over.
S/C	Hello Houston CapCom, Hello Houston CapCom, I expect this is er Brian on the line. This is W5LFL, Spacecraft Columbia, go ahead, Brian. (Houston CapCom uplink is not audible on tape)
S/C	Roger, outstanding. Load and clear. One of the best stations we've heard ever since we've been in orbit. The attitude of the spacecraft is just perfect, of course. We're looking right straight down over Melbourne at just about this moment, this very moment, and we just wanted to establish the fact that we could maintain a backup comm here. John is right behind me and he's been looking forward to er saying hello to you all or having us say hello to you through the backup ham system. And he's giving us a thumbs up signal right now, Brian, so a fine business, we appreciate the chance to talk with you, go ahead, over.
	(Houston CapCom uplink is not audible on tape)
S/C	Yeah, John doesn't want to get on the loop here, I would have to change the headset with him Brian, but hello to you also Bill and why don't we go back now to Australia and take a few moments here before we lose contact line of sight with Joe Kerwin. Thanks a lot, fellas in Houston and back to you Joe, in Australia, over.

JK   OK, thanks very much Owen, I've got the Ambassador, Robert Nesen, here and incidentally standing behind him is Senator Jake Garn from Utah who says hello to John and he still wants a ride. Here's the ambassador.

S/C   OK, fine Joe, thank you, I just passed along the hello from the senator to John who acknowledges that, and er, I am pleased to have you there also Mr. Ambassador so I'm glad to see you all. The Spacecraft attitude really makes the comm system fine here this morning and we've had a lot of good views of Australia, and er, we want to er as I say a moment ago, wish you all er a lot of many thanks and 73's for all the good work that you've done for us here in the past tracking missions. Go ahead."

ORR   Owen this is the ambassador and we welcome you over Australia, glad to have you here, and I'm glad you got the message to John Young from the Senator. He says he's ready to go. Over.

S/C   Alright Victor Kilo One Oscar Robert Robert this is W5LFL spacecraft Columbia and this is one of the best opportunities for communication we've had because the spacecraft is looking right straight down right above you and passed right overhead and we really have had a fine time having a chat and having a conversation with you so we'll (garble) this morning (garble) over the US in about an hour and a half (garble) and we'll look forward to seeing you back home in the United States in another few (garble).

S/C   This is W5LFL spacecraft Columbia saying 73s to all the group there at Orroral and standing by for your final, over.

ORR   Roger, Thanks Owen. We would like to work you again on the same circuit if possible on orbit 126, orbit 126. Do you copy? Over.

       (Background voice) "I think he's gone. That's about it."

       A short while later on orbit 111, during contact through the TDRS, Houston CapCom spoke with John Young on Columbia -

CAPCOM   "Roger John. We wanted to take a few minutes while you guys are all up and give you a little tag-up, and let me just start by just saying that Owen's HAM radio communication over Australia was a big success, well received here and we extend our congratulations to you."

The amateur radio contact with Columbia received considerable publicity within Australia. The experiment was reported in The Canberra Times of 7 December 1983 and was the featured cover story of the magazine Electronics Today International in March 1984.

Left: The interior spread of the article in Electronics Today international.

The cover of Electronics Today International

The Canberra Times article

ACT radio hams talk to Columbia in space

By KAREN MILLINER

Ten ham radio amateurs from the Orroral Valley Tracking Station have succeeded in an experiment to see if emergency back-up communications could be arranged between the space shuttle Columbia and mission controllers in Houston, Texas.

On Monday night, from an amateur radio station in the Deakin Switching Centre, built in their spare time, they contacted astronaut and fellow ham radio enthusiast Dr Owen Garriott on board Columbia as it passed over Australia. They connected him to the Johnson Space Centre in Houston on an international telephone line.

Dr Garriott's conversations were short, as the shuttle passover lasts for only about six minutes, but the voices were loud and clear.

"Glad to see you all," Dr Garriott said. "The spacecraft at-

titude really makes the comm[unication] system fine here this morning. We've had a lot of good views of Australia this morning."

"...This is really one of the best opportunities for communication we've had because the spacecraft is looking right straight down on Melbourne."

Listening in at the Deakin switching centre were the US Ambassador, Mr Robert Nesen, Senator Jake Garn, of Utah, a member of the National Aeronautics and Space Administration appropriations committee, and NASA's representative in Australia, Dr Joseph Kerwin, who is a former astronaut.

Dr Garriott proposed the emergency voice communications test to Dr Kerwin.

He reported that the station was the best he had heard since the shuttle had been in orbit

Chapter Ten - The Final Orbit

In 1979 word came that the planned Tracking Data Relay Satellite System (TDRSS) would take over much of the role of the ground based STDN stations. This space-based system would be capable of tracking earth-orbiting spacecraft with altitudes of less than about 5000 Km. They would be able to provide near 100% orbital coverage compared to about 20% to 25% for the ground stations. In the late 1970s the cost for NASA to run the ground-based tracking stations was over $118 million per year, and escalating at more than twice the rate of most other parts of the NASA budget. In 1977 NASA documents showed the cost of operating Orroral Valley Space Tracking Station was more than that of both Honeysuckle Creek and Tidbinbilla combined. Over time, the TDRSS would save NASA a considerable amount of money.

Thus, the demise of Orroral and most other STDN ground stations like it was imminent, although still some time away. Also, at this time, the support of those spacecraft still using the 136 MHz to 138 MHz VHF frequencies and the 400 MHz UHF frequencies was to be phased out. Supposedly a firm date for the closure of Orroral could not be given until the TDRS system was tested and operational. The first of the satellites, TDRS East, was positioned over the Atlantic Ocean, near South America, to take over an area covering approximately from Hawaii in the Pacific Ocean eastwards to the Seychelles in the Indian Ocean. Full takeover of tracking from space could not occur until there were at least two of the TDRS spacecraft operating.

By late 1983 it was known that the planned date of closure for Orroral was to be 31 January 1985, and that actual tracking operations would cease on 21 December 1984. According to NASA planning documents, the phasedown of the STDN stations was to occur after the TDRSS became fully operational in 1984. However, this planned date was delayed by the tragic 'Challenger' disaster, in which the second TDRSS satellite was destroyed.

In line with NASA requirements to phase out the STDN tracking stations, by early 1982 Orroral had started to decrease operational and other support staffing. This led to shift work being reduced. The first step was removing the night shift, the shift from midnight to eight AM local time. Goddard Spaceflight Centre was advised that tracking requirements now had to be scheduled to meet with the reduced staffing at Orroral.

In early 1983, further reductions in staffing meant that the station was now only able to support one pass (track) at a time, although it was able to resume 24-hour support again with this restriction. Under this arrangement, each tracking shift was reduced to just eight people, plus a powerhouse operator and a guard. The communications centre was not staffed for the evening and night shifts, but there were special arrangements to handle urgent messages. For the evening and night shifts the cooks prepared pre-ordered meals and these were left in the canteen refrigerator for staff to heat as required.

A further reduction in operational staffing took place in August 1984. This reduced the number of operations staff on each shift to six. Any unforeseen staff absences had a severe impact on the support the station could provide, and occasionally resulted in Goddard Space Flight Centre being advised that Orroral could not support some scientific spacecraft tracks.

The new Orroral canteen shortly after construction in 1969.

Part of the dining area in the canteen.

```
GNE140A
RR ANBE GSTS HMSC GARD GUNV
DE GCEN 095B
30/23002
FM NOCC
TO CAN/STADIR OPSR
INFO HMSC/J CONDITT/OPS PLANNER & WHITEHURST/SCHEDULING HEINZ
GUNV/P'C HENRY
GSTS/COMMGR
GARO/RYAN SMITH

OPN M2018 STS-51A
SUBJECT: STS-51A MISSION TIMELINE

1. THE FOLLOWING TIME LINE IS BASED UPON A LIFTOFF OF 1318Z
   7 NOV 1984.

2. INTENDED TO BE USED FOR PLANNING PURPOSES ONLY.

3. THE OFFICIAL SCHEDULE WILL COME FROM NOCC SCHEDULING.


                                  FIRST       LAST
       DATE      H-45     H-30    AOS         AOS        RLS
       NOV 7              NO SUPPORT
       NOV 8     0717     0732    0802        1112       1228
       NOV 9     0551     0606    0636        1122       1235
       NOV 10    0603     0618    0548        0959       1115
       NOV 11    0444     0459    0529        1015       1132
       NOV 12    0334     0349    0419        0906       1025
       NOV 13    2357     0412    0442        0932       1048
       NOV 14    0249     0304    2334        0822       0939
       NOV 15    0315     0330    0400        0851       1005

4. YOUR STATION IS SCHEDULED TO SUPPORT THE FOLLOWING ORBITS:
       DATE            ORBIT NO.
       NOV 7           NONE
       NOV 8           13,14,15,
       NOV 9           28,29,30,31
       NOV 10          44,45,46
       NOV 11          59,60,61,62
       NOV 12          74,75,76,77
       NOV 13          90,91,92,93
       NOV 14          105,106,107,108
       NOV 15          121,122,123,124

SHUTTLE NETWORK OPERATIONS MANAGER SENDS

30/2303Z OCT 84 GCEN
```

The details for the tracking of Discovery on the final flight tracked at Orroral

A 'Tiger Team' was still used to supplement the shift for space shuttle tracking, but STS-51A during November of 1984 was the last gasp for manned spaceflight support at the Orroral tracking Station.

The final orbit of manned spaceflight tracking at the Orroral Valley Space Tracking Station took place during the 14th Space Shuttle flight. This was tracking the orbiter 'Discovery' on orbit 124 of the STS-51A flight, on 16 November 1984. The first orbit of this flight that was visible at Orroral was orbit 13. STS-51A was a 28.5-degree inclination orbit and provided low elevation passes at Orroral. Low elevation passes were always difficult to track because of the effect of the mountainous horizon on the radio signals. Orroral was scheduled to support tracking on 28 orbits of this flight. The tracking orbits occurred in groups of three or four.

The author operating the DDPS console at the 642B computers, November 1984.

With the end of the STS-51A flight the tracking shifts had a few more passes of other scientific satellites before tracking support ended on 21 December 1984. Then, with their work complete, the final shift left the station. There was a small operations crew, including the author, which remained until 31 January 1985, to finalise operational and tracking matters before they too left for good. It was these people who finally and sadly 'turned off' the tracking lights at the Orroral Valley Space Tracking Station.

Due to some problems which had been experienced with the closure of stations in other parts of the world, both administrative and political, it seemed that the USA and Australian governments wanted the closure of Orroral Valley to be as low key as possible. Unfortunately, therefore, Orroral Valley Space Tracking station did not come to a dignified and graceful end. There was no closing ceremony or formal end to the station. The staffing had been diminished over time to the point where there was virtually only a skeleton crew remaining.

People just walked out the door and left it behind, many were simply too overcome with emotion to even look back. Even in later years, some of the former Shiftys would not visit the old site, because of the poignant memories it brought back.

```
ORR001A
RR DSDN DSDM DSSW GIPD GTWL GOPS HMSC
DE AORR 001
21/0001Z
FM OPSR
TO ALL

SUBJ: VALE! ORRORAL!

IT IS WITH REGRET WE MOURN THE PASSING OF ORRORAL FROM THE SPACE
TRACKING NETWORK. AFTER ALMOST 20 YEARS OF SUPPORT, THE FEW OF US
REMAINING (10 LITTLE INDIANS - 9 LITTLE INDIANS - - - - -) LOOK
BACK ON THE MEMORIES AND ACQUAINTANCES GAINED IN THAT TIME.

IN SPITE OF SNOW, FLOODS AND FIRE, ORRORAL HAS MAINTAINED THE
TRADITIONS OF THE NETWORK THROUGHOUT.

WE EXTEND OUR BEST WISHES TO ALL OF OUR FRIENDS AT GODDARD, HOUSTON,
STATIONS AROUND THE WORLD AND TO THE NEW CONSOLIDATED NETWORK.
TO THE PROJECTS PRESENT AND FUTURE, OUR HOPES FOR MORE AND GREATER
SUCCESSES.

FROM THE STATION IN THE PICTURESQUE ORRORAL VALLEY, AUSTRALIAN
CAPITAL TERRITORY, FAREWELL!

A MERRY CHRISTMAS AND HAPPY NEW YEAR TO ALL.

P.S. WE STILL SPEAK "STADAN"!!

21/0002Z DEC 84 AORR
```

The site and buildings were handed back to the Australian Government after closure of the station. Unfortunately, nothing was done to preserve the buildings or the site. This resulted in the buildings and any remaining equipment being devastatingly vandalised. In June 1992 a lone piper (John Wombey) played a final farewell to the vandalised and empty shell of memories left alone and forlorn in the Orroral Valley.

The vandalism was so severe that later in 1992 the remaining buildings were bulldozed, leaving only the foundations.

AUSCI AA62484
NASACOM AA62056

PLEASE PASS TO RONALD S GOLEBY

0/6010/21

PP ANRC AOFF
DE GSTS D24F
AOFF-T-1
ANRC-T-2

PP GFAL
DE GSRM 137B
21/1425Z DEC 84

VZCZCEHB754
PTTUZYUW RUEANATD371 3562112-UUUU--RUWOHEA.
ZNR UUUUU
P 211424Z DEC 84
FM NASA HQ
TO RUWOHEA/RONALD S. GOLEBY DEPT. OF SCIENCE AND TECHNOLOGY BELCONNEN
AUSTRALIA
INFO RUWOHEA/J. MICHAEL STEVENS NASA REPRESENTATIVE CANBERRA AUSTRALI
A
RUWOHEA/GSFC/ 100/ HINNERS, 500/BRACKEN, 530/SPEARING
ZEN/NASAHQS/LI/MORRISON
ACCT SA-XEWA
BT
UNCLAS TN/371
MSG TN-371
FOR ALMOST TWO DECADES THE ORRORAL VALLEY TRACKING STATION HAS BEEN
AN INTEGRAL AND INVALUABLE MEMBER OF NASAS TRACKING AND DATA
ACQUISITION TEAM. WITH THE TRANSFER OF ORR SUPPORT TO CANBERRA AND
THE TERMINATION OF ORR OPERATIONS DECEMBER 21, 1984, I WISH TO
EXPRESS MY DEEP APPRECIATION FOR THE EXCELLENT SUPPORT THAT THE
STATION HAS PROVIDED.
SINCE ORRORAL VALLEY BECAME OPERATIONAL AS A STADAN STATION IN 1965,
THE STATION HAS SUPPORTED A VARIETY OF NASA AND INTERNATIONAL SPACE
PROGRAMS INCLUDING NIMBUS, ISEE, ASTP, OAO, SKYLAB REENTRY, ALSEP,
AND SHUTTLE. THIS SUPPORT HAS BEEN OF THE HIGHEST CALIBER. AS WE ARE
RAPIDLY ENTERING THE TDRSS ERA, THE ADDITION OF ORRORAL VALLEY TO THE
LIST OF PREVIOUSLY CLOSED-OUT STATIONS (WINKFIELD, ALASKA, ROSMAN,
QUITO, BUCKHORN) AND THE CONSOLIDATION OF STDN AND DSN FACILITIES AT
CANBERRA MARK ANOTHER STEP IN THE EVOLUTION OF NASA TD AND A SUPPORT.
IT HAS BEEN A GREAT PLEASURE HAVING THE ORRORAL STATION AS PART OF
THE STDN AND WILL MISS THE EXCELLENT SUPPORT SHE HAS GIVEN. I AM SURE
THAT THIS TRADITION OF EXCELLENCE WILL BE CARRIED OVER TO THE
CANBERRA SUPPORT POSTURE. I WISH EACH MEMBER OF THE ORR STAFF THE
BEST OF SUCCESS IN THE FUTURE AND MY PERSONAL THANKS FOR A JOB WELL
DONE.
S/ROBERT O ALLER
    ASSOCIATE ADMINISTRATOR FOR
    SPACE TRACKING AND EATA SYSTEMS
XXXXX
BT

21/1511Z DEC 84 GSTS

NASACOM AA62056*
AUSCI AA62484

Historic space stations to be demolished within days

Dozens of computers litter the main control room at Orroral.

Pictures: ANDREW CAMPBELL

Orroral tracking station: an eerie testament to history

Tracking stations loss 'a shame'

Pictures: CHARLOTTE FENTON

John Charlton, who has organised a reunion of former tracking station workers: the staff were very close; almost extended family.

Vandals have left their mark.

Newspaper clippings from the time

Eventually, after representations to the Australian Capital Territory government, the Orroral Valley Tracking Station site was determined to be part of the Australian Capital Territory's and Australia's heritage and it was heritage listed in February 2016 to protect what little remains.

The ACT Parks and Conservation service has erected historical markers at some of the significant foundations so that visitors can have some idea of what was once there.

Pat Lynch's prophetic final cartoon. "Who said the station wouldn't fall apart when I left......"

Tony Pelling -

"I think it's sad that a lot of areas of government have never really treasured their own history here, so many things were done at this place. And really a lot of people ended up, who were in the industry here, going over to America and having a really big impact at JPL or Goldstone."

Ken Strickland -

"I just loved it all. I loved it all for 30 years, I just kept going until I retired. I didn't like working in recorders, I can tell you that, but apart from that I just loved the whole thing."

Dave Brown –

"The 10 years that I spent at Orroral Valley were the best working years of my life. I had spent 15 years in the air force and then another 16 years in the public service. Those 10 years at Orroral were the best years, the most interesting years, the best companionship, the best camaraderie between people than anywhere else that I have ever been."

Loretta Edgar –

"Only that we were very much the forgotten station. I don't think people really understand what was involved in the operation of the tracking station and the way it worked. Even what its primary aim was. You talk to people about tracking satellites now, and it's old hat. I don't think people understand the mechanics of it the way it was done and why it was done that way, and the capabilities. The capability of Orroral to track 6 or 7 satellites at once, that was a pretty big deal really."

Author **Philip Clark** -

"From my own point of view, I can say that it was the best job I ever had. For nearly 20 years I never tired of it. There was something happening nearly every day and new things to learn all of the time. The people, the challenges, the learning, the experiences, the environment, it was simply just the best. We were working at the cutting edge of the technology of the time.

My only regret is that it did not last longer. I came from another state on a short contract to start with, and stayed until the close of the station, it was that good. I can only repeat what others have said, and that is Orroral's greatest strengths were the camaraderie of the people and the quality of the work done. I have never experienced anything like it before or since. And it is still present to this day when Orroral people meet in everyday life."

Vale, Orroral Valley Space Tracking Station – The greatest and best of them all!

Photo and Graphic Credits

Many of the photos used in this book have been widely circulated over the years and the actual origins of these have become blurred. This was shown by a number of my own pictures that were submitted to me by others, some in a slightly modified form, but said to be from a different source. For this reason, attribution has only been given where the originator could be determined with reasonable assurance. Where no attribution is given, the origin is unknown or uncertain. Therefore, there may be errors and omissions of attribution for which the author apologises but disclaims all liability for any such errors or omissions.

Image Credits:

Page number	Image	Photographer / attribution
Title page		NASA
iv	Going to work in the snow	Philip Clark.
viii	Map image	Created by Robina Gugler modified by Kim Lambert.
xii	Minitrack antenna array	unknown
xiii	Minitrack chart recorders	unknown
xiii	Minitrack control console	unknown
xiii	Sputnik	Smithsonian Air and Space Museum.
xiv	DSS 41 Woomera	NASA
xiv	World locations map	NASA
xv	Article from Adelaide advertiser	NLA Archives.
1	Photos of Muchea Tracking Station	State Records Office of Western Australia
2	Carnarvon tracking station	Hamish Lindsay

Page number	Image	Photographer / attribution
2	Orroral sign	Unknown
3	STADAN Map	NASA
5	Orroral panorama	Philip Clark.
6	Antenna	Unknown
7	Antennas	Unknown
8	Naas bridge and Mt. Tennent road	Unknown
9	Station car fleet	CDSCC Archives
9	Long Boom Cherry Picker	R Murray
9	Smaller cherry-picker	A Bourne
10	A shift being driver over flooded Rocky crossing	Unknown
10	Rocky Crossing in flood	Unknown
11	Workshop, refuelling station and generators	Unknown
12	Settling pond	Unknown
12	Old canteen	Unknown
12	New canteen	Philip Clark
13	Firefighting training x 2	Unknown
14	Ops building map	Unknown
15	Ops room images x 3	Unknown
16	Crossbar equipment, Telemetry system no 1 pictures and Dynatronics	R Murray
16	Radiation PCM-DHE	Philip Clark
17	Tone command encoder	R Murray
19	printout	NASA
19	Antenna patterns	ARRL
21	WRESAT	unknown
22	Switching centre at Deakin	Canberra Times
22	Voice centre at Goddard Space Flight centre	NASA
24	Rebuilding and extending the ops room 1974 4 images	Unknown
25	Ops room plan	Philip Clark
25	Equipment operator Ed Maly loading tape onto a B & H 8007 data recorder.	Philip Clark.
26	Two Manned Space Flight Telemetry Processors type 2 (MSFTP-2) in the data handling area at Orroral Valley	Unknown

Page number	Image	Photographer / attribution
27	The 9 metre Antenna	Philip Clark.
27	Two views of Multi-Function Receivers.	Philip Clark.
28	Programmable Patch board	Ross Murray
28	The S-Band spacecraft ranging system at Orroral.	Unknown
28	Operator interface with a computer-based Spacecraft Command Encoder	Philip Clark.
29	The station ComTech position. The manned spaceflight voice control console.	Philip Clark.
30	An Operations Supervisor console	Hamish Lindsay
31	The 642B computers (centre) at Honeysuckle Creek before the move to Orroral.	Hamish Lindsay
31	The 642B computers at Orroral.	Philip Clark.
34	Warning tag and welcome sheet	Philip Clark.
35	Two views of the long-range radar at bermuda	Philip Clark.
35	a view of the Bermuda tracking station	Unknown
36	Course members at Bermuda	Philip Clark.
36	Bermuda map	Unknown
37	Beer mug and course members at Palmetto Bay Hotel	Philip Clark.
39	Soyuz and Apollo launches	NASA
40	Apollo from Soyuz and Soyuz from Apollo	NASA
40	ASTP decal	Philip Clark
41	ATS-6	NASA
42	Cartoon Illustration	Pat Lynch
43	Cartoon Illustration	Pat Lynch
45	Antenna graphic	ARRL
46	Cartoon Illustration	Pat Lynch
46	ASTP Mission Control Houston	NASA
50	Decal	Philip Clark.
51	Letter	Philip Clark.
51	Apollo Soyuz	Unknown
52	Letter	Philip Clark.

Page number	Image	Photographer / attribution
53	Cachet and certificate	Philip Clark.
54	3 images	Philip Clark.
57	Certificates and evaluation report	Philip Clark.
58	DDPS system 2 images	Philip Clark.
59	Yarragadee installation	Philip Clark.
60	Yarragadee antenna installation 3 images	Philip Clark.
61	Calculation sheet	Philip Clark.
64	Peter Uzzell	Unknown
70	Tracking schedule	Lindsay Richmond
71	Diagram	Philip Clark.
72	Multifunction receivers, and Rob Quick at DDPS console	R Murray
72	MSFTP-2	Unknown
73	Two images	Philip Clark.
75	Orbit map	Philip Clark.
76	Shuttle take-off and landing images	NASA
83	3 images of presentation	Philip Clark.
84	Shuttle cake, meal and flag	Philip Clark.
85	Decal collection	Philip Clark.
87	Press release	NASA
88	Joe Kerwin at Deakin	Unknown
88	Antenna at Deakin building – two images	Philip Clark.
89	Note re Amateur radio experiment	Richard Elliott
90	Schedule	Richard Elliott
91	3 images – set up of equipment	Philip Clark.
92	Configuration diagram and Dr Garriott radio contact card	Philip Clark.
93	Track map	NASA
93	Author's radio contact card	Philip Clark.
95	Interior Electronics Today international	Philip Clark.
96	Cover ETI and Canberra Times article	ETI and Canberra Times
98	Canteen after construction	Philip Clark.

Page number	Image	Photographer / attribution
98	Tracking schedule	NASA
98	Canteen interior	R Murray
99	Author at DDPS console	R Simons
100	Message	Ian Grant
100	Two images at abandoned station	Unknown
101	Telex message	Ian Grant
102	Newspaper clippings	Ian Grant
103	Cartoon	Pat Lynch
119 - 134	All images of documents	Philip Clark.
138-139	All images of documents	Philip Clark.
141-147	All images of documents	Philip Clark.

About the Author

Philip Clark came to Orroral Valley Space Tracking Station at the end of 1966 as a technician. When the station closed in 1985 he was the Senior Operations Supervisor (SOS).

After leaving the Spacetracking industry, he was engaged in senior technical, engineering and research positions in government and private industry. In 1993 he was awarded a National Medal of Australia for Service. In 1999 he gained a Master of Science degree from the University of New South Wales at the Australian Defence Force Academy in Canberra.

He maintains an interest in amateur radio, through which he has been one of few people to speak both with Russian Cosmonauts on the USSR 'MIR' space station and USA astronauts on the Space Shuttle 'Columbia' from his car! He is the author of a number of books, technical manuals, and of articles for technical magazines. At the time of writing he is retired.

Other Books by Philip Clark

Philip's first book, Acquisition, is currently out of print, but a new edition is being planned.

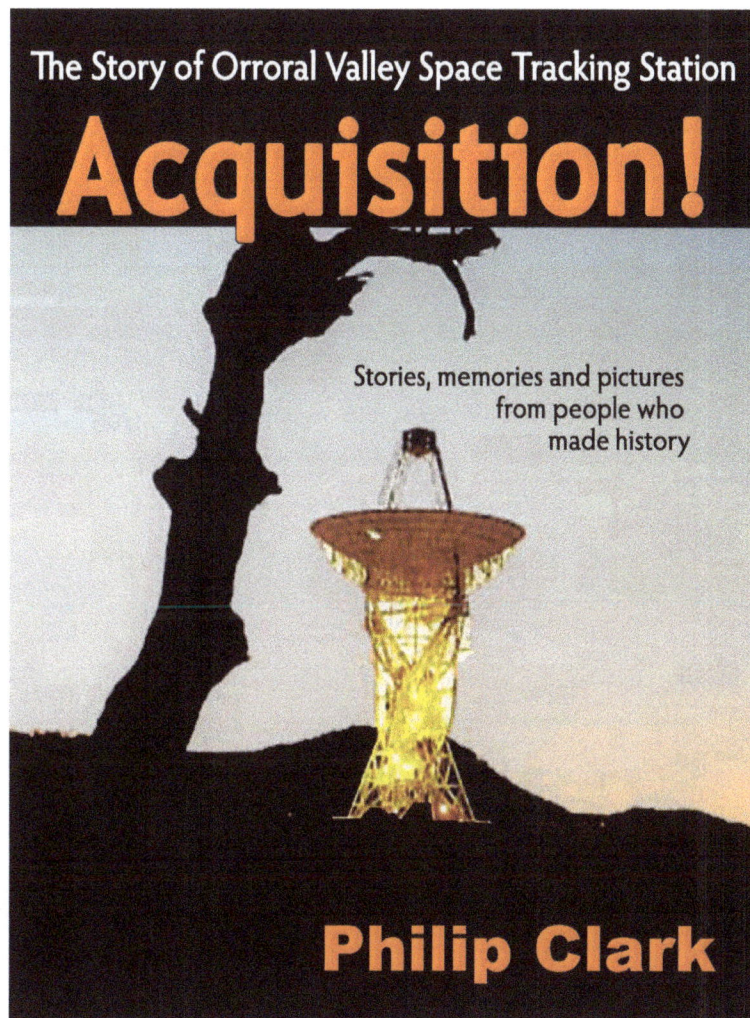

Bibliography

AWA, 'Register of Shift Change Notices 3.3 1977 – 1984', Orroral Valley Tracking Station internal document, 1977

AWA, 'STS-2 M2002 Local Operations Procedure', Orroral Valley Tracking Station internal document, 1981.

Baker, David, 'Space Shuttle', New Cavendish Books, 1979.

Clark, Philip, 'Acquisition – The Story of Orroral Valley Space Tracking Station', Philip Clark, Canberra, 2012

Coates, R. J, 'Tracking and Data Acquisition for Space Exploration' In Journal: Space Science Reviews, Volume 9, Issue 3, pp. 361-418

Cortwright, Edgar M. (Ed), 'Apollo Expeditions to the Moon;, NASA, SP-350, Washington DC, USA, 1975.

Dench, Paul and Gregg, Allison, 'Carnarvon and Apollo One giant leap for a small Australian Town', Rosenberg Publishing Pty Ltd, Dural, N.S.W. 2010

Dougherty, Kerrie; James, Matthew L., 'Space Australia- The Story of Australia's Involvement in Space', Powerhouse Publishing, Haymarket, NSW, Australia, 2000.

Dougherty, Kerrie, and Sarkissian, John, 'Dishing up the Data: the role of Australian space tracking and radio astronomy facilities in the exploration of the Solar System' Powerhouse Museum, Sydney, Australia, and CSIRO, ATNF-Parkes Observatory Parkes, Australia.

Ezell, Edward Clinton and Ezell, Linda Neuman, 'THE PARTNERSHIP - A History of the Apollo-Soyuz Test Project', National Aeronautics and Space Administration, 1978

Froehlich, Walter, 'Apollo Soyuz', NASA, EP-109, Washington, USA, 1976

Hall, Gerald (Ed), 'The ARRL Antenna Book', The American Radio Relay League, Newington, Connecticut, USA, 1988

Joels, Kerry Mark and Kennedy, Gregory P. 'The Space Shuttle Operator's Manual', Ballentine Books, 1982

Lindsay, Hamish, Essay 'Muchea Tracking Station', www.honeysucklecreek.net.au, date unknown

Lindsay, Hamish, 'Tracking Apollo to the Moon', Springer-Verlag, London, 2001.

Manned Flight Operations Division, 'An Introduction to The Manned Space Flight Network', GSFC, Greenbelt, Maryland, USA, 1965

NASA, 'Apollo Soyuz Press Kit', NASA, release No. 75-118, 10 June 1975.

NASA, 'NASA Facts - Apollo Soyuz Test Project', NF-52/5-75, 1975.

Tsiao, Sunny, 'Read You Load and Clear', NASA History Division, NASA SP-2007-4232, Washington, USA, 2008

Acronyms and Abbreviations

ACT	Australian Capital Territory
ADFA	Australian Defence Force Academy
ALSEP	Apollo Lunar Surface Experiment Package
AM	Amplitude Modulation
AOS	Acquisition Of Signal
ARIA	Apollo Range Instrumentation Aircraft (A tracking aircraft)
AWA	Amalgamated Wireless Australia Ltd. An Australian electronics company
CAPCOM	The person who talks to the astronauts on manned space flights. Usually another astronaut
CDSCC	Canberra Deep Space Communications Complex. (Tidbinbilla tracking station ACT)
Cherry-picker	A truck with a working basket mounted on an extendable hydraulically operated arm
Col Tower	A collimation tower used for testing and calibrating antennas.
COMTECH	Communications Technician. Person responsible for the technical and operational aspects of voice communications to manned spacecraft.
DDPS	Digital Data Processing System
DFE	Callsign of Data Flow Engineer (Houston)
DHE	Data Handling Equipment.
DFI	Development Flight Instrumentation
DHS	Data Handling System
DLSM	Data Link Summary Message
DOS	Department Of Supply, Australian government
DSN	Deep Space Network
DSS	Deep Space Station
EMI	An electronics company
EC	Engineering change document
FET	Field Effect Transistor
FM	Frequency Modulation

CAM	Computer Address Matrix
GaAs	Gallium Arsenide
GC	Ground Controller (Houston)
GMT	Greenwich Mean Time (see also UTC)
GPS	Global Positioning System. A satellite network of precision timing signals that allows a user to determine their position to within less than a meter
GSFC	Goddard Space Flight Centre
HAW	Hawaii STDN Station
HF	High Frequency radio
Hi-Ranger	Manufacturer's brand name (See 'Cherry-Picker')
IGY	International Geophysical Year (1 July 1957 - 31 December 1958)
JPL	Jet Propulsion Laboratory
KBPS	Kilobits per second
Kg	Kilogram
KHz	Kilohertz - measurement of radio frequency
Kw	Kilowatt
LASER	Light Amplification by Stimulated Emission of Radiation
LED	Light Emitting Diode
LNA	Low Noise Amplifier
LOP	Local Operations Procedure document
LOS	Loss of Signal
M&O	Maintenance & Operations supervisor
MCC	Mission Control Centre (Houston)
MFR	Multifunction Receiver
MHz	Megahertz - measurement of radio frequency
MILA	Callsign of Merritt Island Tracking Station, Florida USA
MSFN	Manned Space Flight Network
MSFTP	Manned Space Flight Telemetry Processor
MUX	Signal multiplexing unit
NSW	New South Wales
NASA	National Aeronautics and Space Administration
NASCOM	The NASA world-wide network of communications for space tracking operations
Neg Acq	Negative Acquisition of signal. Nothing received
NOCC	Network Operations Control Centre (GSFC)

NOM	Network Operations Manager
NOSP	Network Operations Support Plan (See also Opsplan)
NST	Network Support Team
OD	Operational Data
OE	Operations Engineer. (See also SOS)
Operations Coordinator	Person in charge of a section of the operations area at Orroral.
OPSCON	Operations control Centre at Goddard Space Flight Centre
OPSPLAN	Operations Plan (Later called NOSP)
OPSR	Operations Supervisor, a special position for Space Shuttle passes.
ORR	Orroral Valley Tracking Station callsign.
PASS	The length or duration for the tracking operation of a satellite in view of the station.
PCA	Point of closest approach of a satellite to the station.
PCM	Pulse Code Modulation
PM	Phase Modulation
QUINDAR	Company name of a system used over voice circuits to remotely switch equipment such as transmitters. Makes the 'beep' sounds heard during communication with astronauts.
QUITO	Callsign of Quito STDN station Ecuador
RF	Radio Frequency
RTC	Callsign of Real Time Command Controller (Houston)
RTD	Real Time Data. Data showing what is happening now.
SAO	Smithsonian Astrophysical Observatory
SATAN	Satellite Automatic Tracking Antenna Network
S-Band	Frequencies in the 1700 MHz to 2300 MHz range.
SCAMA	Station Conferencing And Monitoring Arrangement. The voice communications link between all NASA facilities and Goddard Space Flight Centre.
SCE	Spacecraft Command Encoder
SCM	Site Configuration Message
SCVM	Shuttle Command & Voice Multiplexer
Servo	Combination Electronic and hydraulic system for controlling and positioning antennas
SITE COORD	Station Coordination voice communications circuit
SOS	Senior Operations Supervisor. (See also OE)
SRE	S-Band Ranging Equipment
SRT	System Readiness Test
SSM	Site Status Message
STADAN	Space Tracking And Data Acquisition Network

STADIR	Station Director
STDN	Space Tracking Data Network
STS	Space or Shuttle Transportation System
SYNC	Synchronisation
TDRS	Tracking Data Relay Satellite
TDRSS	Tracking Data Relay Satellite System
Tech	Abbreviation for technician
TID	Tidbinbilla tracking station callsign
TLM	Telemetry
TRACK	Callsign of Houston Tracking Controller
UHF	Ultra-High Frequency (Above 300 MHz)
UNSW	University of New South Wales
US	United States
USA	United States of America
USB	Unified S-Band microwave radio system
USSR	Union of Soviet Socialist Republics
UTC	Universal coordinated time (See GMT)
VHF	Very High Frequency (50 MHz – 300 MHz)
WRE	Weapons Research Establishment. An Australian government establishment once part of the Department of Supply and initially at Salisbury South Australia. Now Defence Science and Technology Organisation (DSTO)
WRESAT	Weapons Research Establishment Satellite. Built by WRE and University of Adelaide.
Yagi	A type of antenna named after Prof Yagi of Japan
Z	Zulu time zone (GMT or UTC)

APPENDICES

Appendix 1 - Station Closure Documents

DEPARTMENT OF SCIENCE AND TECHNOLOGY
Entrance 5, Benjamin Offices, BELCONNEN TOWN CENTRE

-1 JUN 1981

ORRORAL VALLEY

Postal Address
P O Box 65
BELCONNEN ACT 2616

Telegrams: CONSCIENCE   Telex: AA62484

ADDRESS ALL CORRESPONDENCE TO THE SECRETARY

REFERENCE   79/1905

CONTACT OFFICER   R.A. Leslie

TEL.   64 4080

Mr C. McDonald,
Secretary,
Trades and Labour Council
  of the ACT Inc.,
P.O. Box 119,
MANUKA, A.C.T.   2603

Dear Sir,

Space Tracking Industry

        I refer to previous advice and discussions on
NASA's long term plans for space tracking activities in
Australia.

        You will recall that because of the diminishing
number of earth orbiting satellites dependent upon ground
tracking support, the future role of the Tracking and Data
Relay Satellite System and the changing requirements for the
support of future deep space missions, NASA had decided to
consolidate its space tracking networks into a single network
under the control of the Jet Propulsion Laboratory (JPL). The
goal is to complete the consolidation in 1985 so as to be
ready to support the Voyager/Uranus encounter in November 1985
in the new mode. It was explained that planning and studies
were continuing and that we would keep you informed of
developments.

        We had further discussions with NASA earlier this
month on the long term plans to consolidate the networks and
the main points were:

/2

2.

(a) NASA confirmed that it was still its plan that the networks be consolidated by 1985 but pointed out that the plan would be subject to further development and refinement.

(b) Because a study had shown that significant technical advantages could be gained by locating all antennas at a common site, NASA advised that it was its intention to locate all antennas at Tidbinbilla. This will result in the eventual closure of the Honeysuckle Creek and Orroral Valley Tracking Stations probably about 1985.

(c) A proposal that new antennas be provided rather than re-locate existing ones is being studied. However this would not mean that the existing antennas at Honeysuckle Creek and Orroral Valley would be retained.

(d) A study is being made of requirements for support of satellites that were not included in the original plans for the consolidation of the networks. There are many options as to how they may be supported in the short term, but the choice will not affect the long term future of the stations.

(e) When all tracking activities have been re-located at Tidbinbilla it is expected that the staffing requirement will be about half the present level at the tracking stations. We are confident that the diminishing workload will allow a good part of the reduction to be achieved, gradually, by 1985, through normal attrition.

(f) NASA will be conducting a further review of the plans for network consolidation in late October.

As you can see from the above more investigation and studies are to be completed before a definite plan can be established. We expect to be able to provide you with further information later in the year after the results of NASA's review in October become known.

/3

3.

    We are prepared to convene a meeting to discuss
the plans as they stand now if that is your wish, but in
view of the uncertainties it may be preferable to await the
outcome of the October review.

    A copy of this letter has been forwarded to the
Secretary/APEA, Sydney, and to the Federal Industrial
Officer/AAESDA, Melbourne, and it would be appreciated if
you would forward copies to the local representatives of
the unions respondent to the Space Tracking Industry Award.

Yours faithfully,

(R.A. LESLIE)
Senior Assistant Secretary
(Space Projects Branch)

29 May 1981

```
GOR050A
RR AORR ANBE
DE GSTS 042
14/2112Z
FM CODE 850/W BODIN/NASA GSFC GREENBELT MD
TO AORR/STADIR
INFO DLD/CODE 855.4/W JONES
ANBE/T REID

SUBJECT  ORR 9 M ANTENNA DEACTIVATION
YOU ARE AUTHORIZED TO DEACTIVATE THE ORRORAL 9 M ANTENNA ON
FEBRUARY 20, 1984 AND PROCEED WITH THE TEAR DOWN AND SHIPMENT
IN ACCORDANCE WITH EC 4028-500S AND EC 4033-500S. RELOCATION
OF THE S-BAND EXCITER AND TEAR DOWN OF THE 9 M PEDESTAL BUILDING
MAY BE DONE AT A LATER DATE WHEN DOWN TIME IS AVAILABLE. WE STILL
REQUIRE THE 9 M PEDESTAL BUILDING TO BE SHIPPED TO VAFB.
W BODIN SENDS.

14/2345Z FEB 84 GSTS
```

1984 FEB 14 23 57

DEPARTMENT OF SCIENCE Minute

THE SECRETARY
THROUGH: FASSSD

PURPOSE

To inform you of the transfer of the Orroral Valley facility to the Department of Territories.

BACKGROUND

2   As a result of the NASA decision to consolidate its operations in Australia the Orroral Valley Tracking Station was closed in January 1985.

3   In anticipation of that event the buildings and facilities at that site were offered to the then Department of Territories and Local Government (DTLG). This offer was accepted in April 1984 by the Minister for that Department.

DISCUSSION

4   Subsequent to the redeployment of the Orroral Valley electronic equipment in early September 1985, the Department of Territories was requested to take responsibility for the site.

5   Agreement to this request was received, with the Department of Territories assuming responsibility for the site from close of business on 13 September 1985. In addition permission was given by Department of Territories for the storage of certain items of surplus equipment at Orroral Valley until disposal action could be undertaken through the Department of Local Government and Administrative Services.

6   In June 1985 a letter was received from DTLG indicating that the Namadgi National Park Consultative Committee was interested in preserving some of the tracking equipment from Orroral Valley.

7   This equipment would form part of a historical and cultural display to be centred in one of the former tracking stations in the Namadgi Park.

8   Accordingly, arrangements are underway to transfer most of the Apollo Data Processing System to Department of Territories along with the Orroral Valley facility. This equipment was used to process data from the Apollo Moon Landing Mission and is thus of great historical interest.

2

RECOMMENDATION

9   That you note the transfer of the Orroral Valley site
to Department of Territories and that assistance is given to
that Department to set up a historical display dealing with the
Apollo project. This assistance would be at no cost to this
Department and would cover the provision of models, photographs
and information pertaining to the Apollo project.

R S Goleby
ASSP

September 1985

Appendix 2 – Staff Changes

INTERNAL MEMORANDUM		ORRORAL VALLEY TRACKING STATION	
TO	CHIEF ENGINEER	YOUR FILE NO.	
FROM	SOS	OUR FILE NO.	3.1
COPIES		DATE	831121

SUBJECT :   CHANGEOVER TO SINGLE LINK OPERATIONS.

In order to implement the changeover to single link operation, the following previously discussed items will require immediate action to ensure that they are resolved prior to the end of December, 1983.

1.   Schedule Change procedures.

As discussed, GSFC will need to vary procedures to ensure that all changes to scheduled support are notified by voice in addition to normal teletype advice.

2.   Teletype Traffic.

As manning of the Communications Centre will be considerably reduced, action must be taken to ensure that operations personnel are aware of relevant teletype traffic as soon as possible. The following steps should be implemented, as discussed, to ensure this.

(a)   Install a R/O printer between the link 1 and link 2 DDPS consoles. This printer would normally be patched to the 'A' receive teletype circuit.

(b)   In consultation with GSFC and Deakin Switch, arrange for all incoming TTY traffic except pointing data and logistics messages, to be sent on the 'A' circuit.

3.   Communications Procedures.

In consultation with GSFC, NASCOM and Deakin Switch, incorporate into standard procedures and documentation, the following:

(a)   Outside of the hours of 0800 - 1630 local, Monday to Friday, and on public holidays, rostered off days and on other days due to staff absences, Orroral Communications Centre will not be staffed for routine requirements.

(b)   During these periods two Scama lines will be extended to the operations area together with the fault-net order wire.

........./2.........

- 2 -

(c) Whenever possible, teletype machines will be checked to ensure that paper and ribbons are adequate and there are no jams. Messages will not be distributed until normal manning is resumed.

(d) Except for contingencies outside of normal manning hours, there will be no routine line checks, number comparisons or configuration changes during this period.

(e) Administrative procedures to advise of this state (i.e., messages, notifications, etc.,) should be kept to an absolute minimum.

4. The logistics terminal and teletype circuit should be relocated to Tidbinbilla.

5. One of the extension 28 telephones presently located in the Shift Supervisor's Office should be relocated to near the link console area, so that night-switched calls can be answered.

6. Formulate and distribute to supervisors the notification and support procedures to be used when insufficient staff report for duty to enable proper manning for support.

P.G. CLARK, SOS.

INTERNAL MEMORANDUM   ORRORAL VALLEY TRACKING STATION

TO	SHIFT SUPERVISORS A, B, C, D	YOUR FILE NO.	
FROM	S.O.S.	OUR FILE NO.	3.6
COPIES	CE SD (Tid) Orroral Guard SSA H. Westwood House Front Office W.I.S.	DATE 23 January 1984	

SUBJECT :   SHIFT TRANSPORT

The operational and powerhouse staff have been combined onto four
shifts. Near coincidence between these shifts and guards shifts
has been provided to minimise transport rearrangements as much as
possible. This memo outlines the procedures to be used for the
provision of shift transport. The A, B, C, D teams as at 840101
are:

A.	B.	C.	D.
G. Owttrim	H. Cocking	I. Cockburn	H. Parker
K. Hills	B. Eagleton	R. Quick	I. MacAndrews
W. Mann	D. Pritchard	J. Valenzuela	D. Fallow
L. Richmond	G. Hobbs	K. Strickland	D. Hickson
J. Gales	N. Murphy	J. Kerr	E. Maly
A. Sholtez	C. Steele	C. Guest	S. Gordon-Douglas
M. Barritt-Eyles	D. Bailey	R. French	J. Charlton
J. Quinones	R. Henry	B. Lynch	A. Penney
J. Balthazaar	G. McEwan	H. Galea	A. Payne

The responsibility for shift transport rests with the A, B, C and D
shift supervisors.

There will be 3 cars normally available to transport each shift.
Spare cars may be used as required.

Shift cars allocated for departure at 1600 Monday, Thursday and
Friday will be left in the marked shift bays. Keys will be left
in the vehicles and will not be delivered to the Shift Supervisors
office. Cars allocated to guards at 1600 on Monday will be retained
by them for use on Tuesday and Wednesday. The keys for these
vehicles will be held in the Orroral guard house while they are on
station. The cars will be left in the parking area.

Car moves will be necessary when changes of shift occur.

Night shift (0000 - 0800)

(a)   Friday to Wednesday - arrange transport on a daily basis.

(b)   Thursday - Prepare and distribute a transport list for the
       following evening shift beginning on Tuesday. Ensure that
       nominated drivers receive copies of the list.
       Provide copies to W.I.S. if required to be left in delivered
       cars.

(c)   When departing at 0800 Tuesday or Wednesday, spare cars may be
       used to replace those retained by the guards or not returning to
       station that day. The guards cars are then available as spares
       if required.

.../2

Arrange cars for the new night shift in accordance with the distributed transport list. As there is an overlap with guards at this change, only one car move should normally be required.

Day Shift (0800 - 1600)

(a)   Thursday to Sunday - arrange transport on a daily basis. Provide transport for cooks on weekends, public holidays and other days when normal day transport not available.

(b)   Monday - prepare and distribute a transport list for the night shift neginning on Friday. Ensure that nominated drivers receive copies.

        Provide copies to W.I.S. if required to be left in delivered cars.

        Arrange cars for guards on Tuesday and Wednesday. The Guards will use the same cars on Tuesday and Wednesday and will retain the keys in the Orroral guard house while on station.

Evening Shift (1600 - 2400)

        Arrange transport on a daily basis.
        On the last day of the shift, arrange car moves for the evening shift commencing on Tuesday.
        Prepare and distribute a transport list for the day shift commencing on Thursday.
        Ensure that nominated drivers receive copies and, if applicable, W.I.S.

        Arrange the car moves for the day shift commencing on Thursday. Two cars may finish with operations personnel on Monday, and two will finish with guards on Wednesday. Notify W.I.S. *of* the car moves required and that they are to take place following the end of shift at 1600 on Wednesday.
        Up to four car moves may be required.

P.G. CLARK
S.O.S.

840123

INTERNAL MEMORANDUM   ORRORAL VALLEY TRACKING STATION

TO S/S A,B,C,D,	YOUR FILE NO.
FROM SOS	OUR FILE NO. 3.1
COPIES COM-CEN SD CE SD (TID) H. WESTWOOD	DATE 840806

SUBJECT: OPERATIONS PROCEDURAL CHANGES

Due to the recent reduction in operations staff and the curtailment of the ops clerk position, it is necessary to reduce the workload in the shift Supervisor/operations area.
To effect this reduction, the following procedures will be implemented.

1. The amount of filing required in the shift Supervisor's office will be considerably reduced. The only files to be maintained in this office will now be:
   a) Schedules and Amendments
   b) Briefing messages
   c) ESR
   d) TOR
   e) TSR

One copy of all other messages will be provided for attachment to respective clipboards. ISEE Daily mode schedule handling will remain unchanged.

2. There will be no filing of any pointing data in the shift Supervisor's office. If data is not available on disc then it may be necessary to request any required data.
Special (eg. launch etc). INP's will be filed on the spacecraft file in the SOS office.

3. All spacecraft files and other administrative files will now be maintained in the SOS office. If urgent reference is required to any of these files, the duty shift Supervisor may request the guard to open the SOS office for this purpose.

4. The top (black) copy of all TTY messages, except those listed in para 1., will be delivered to the SOS office to be retained as copy masters.

5. DCN's will be routinely actioned by the SOS office, where immediate implementation of a DCN is necessary, the duty shift Supervisor may action it in the operations copy of the document. The shift Supervisor is not required to produce photo-copies of DCNS.

6. Operations scheduling will be the responsibility of the duty shift Supervisor only. There will not be clerical assistance available for this function. There may be times when shifts commence duty that the operations schedules are not ready. It is accepted that this may, on occasion, result in delayed AOS or even a missed pass.

7. Briefing messages will be retained on the briefing message file and will be copied and distributed by the duty Supervisor when required. There should now be no need to have a number of photo-copies of these retained for the future use.

8. The rotation of the 14 day data bins will be the responsibility of the day shift supervisor or the SOS office when a day shift is not on duty.

9. Teletype distribution instructions will be amended in the near future to meet these requirements.

P.G. CLARK
SOS.

INTERNAL MEMORANDUM		ORRORAL VALLEY TRACKING STATION	
TO S/S A,B,C,D		**YOUR FILE NO.**	
FROM SOS		**OUR FILE NO.**	3.2
COPIES CE H.WESTWOOD SD SD(TID)		**DATE**	840726

SUBJECT: REVISED SHIFT STRUCTURE

A further transfer of staff to Tidbinbilla will take place on the 6th of August. This will reduce the number of operations personnel on each shift to 6. There will be some people transferred between shifts at Orroral to maintain a reasonable balance within each shift.

Due to the minimal staffing levels and in accordance with management directives, further recreation leave will not be granted, apart from exceptional circumstances, until after 30th September.

At each shift handover, supervisors will assess the impact of any staff absences on the scheduled workload and decide if there is a requirement for off-going personnel to be held back for all or part of a shift on an overtime basis. If subsequent changes to the scheduled operations result in insufficient staff being available to meet support requirements, then GSFC must be advised that the station is unable to support all or part of an operation and that the Station Director will be in contact to explain the reason. Contact the Station Director as advised in the memo dated 840217, copy attached, and explain the situation to him so that he can advise GSFC accordingly.

Extended absences of shift personnel will be covered from day staff, where possible.

P.G.CLARK,
SOS

INTERNAL MEMORANDUM   ORRORAL VALLEY TRACKING STATION

TO	Distribution		YOUR FILE NO.	
FROM	SOS		OUR FILE NO. 3. 1/2014	
COPIES	SD	ACT Manager AWA	DATE	
	SD(TID)		840820	
	Front Office	SOS (TID)		

SUBJECT: Shuttle Team Personnel - Time and Personnel adjustments.

The scheduled period for the STS-41D (M2014) mission spans rostered-off or non-working days for most personnel assigned to the shuttle team. These days will be adjusted at the conclusion of the mission in accordance with the following guide lines.

1. Each person on the shuttle team will be allowed to take time off without pay to adjust the rostered-off or non-working days actually worked during STS-41D support. For the purpose of this adjustment only the Saturday and Sunday, 840901 and 840902 for persons normally on daywork, or non-working days for C shift that do not fall on a Saturday or Sunday, will be considered. Penalty payments will be made for Saturday and Sunday work.

2. These days off may be taken any time in the 4 week period following conclusion on the STS-41D support by mutual agreement between the person concerned and their supervisor. The days may be taken singly or consecutively. The time off not taken by the end of the 4 week period will be lost. Supervisors will ensure that suitable transport arrangements are made.

3. In all cases, supervisors will annotate time cards with the remark "Rostered off day-Shuttle" to allow the pay office to account for this time.

4. Those personnel from B shift assigned to the shuttle team will not require shift transport but will be transported with the shuttle team. They will work Tuesday 840904, as overtime; and Wednesday 840905, if the mission is extended. Time off will be adjusted in accordance with the above guide lines if 840905 is worked.

5. The following staff will be assigned to the shifts shown to maintain operational staffing levels during the STS-41D mission period.

   K. Hills - A Shift   B. Lynch - C Shift
   H. Cocking - B Shift   G. Valenzuela - D Shift
   A. Sommariva - B Shift

   Supervisors should arrange transport when shift changes are issued showing transfer dates.

   P.G. CLARK
   SOS

Distribution:
G. Hobbs	S/S A
D. Pritchard	B
D. Hickson	C
K. Strickland	D
I. Barr	CE
R. Henson	SE(DA)
B. Ormeno	▉▉▉▉▉
D. Richards	AWA(SYD)
S. Benning	A. Bailey
E. Cook	J. Wells
L. Hopson	N. Eyre

Appendix 3 - Employment Contract 1965

E.M.I. ELECTRONICS LIMITED

For agreement with Semi-Professional Staff recruited outside
the Australian Capital Territory

Dear Mr. Clark,

This is to confirm the offer of employment at the Data Acquisition Facility, Orroral Valley, A.C.T. made to you by this Company
commencing on .... and terminating on 7.9.1967.

The date on which you will be required to take up permanent
duty in the A.C.T. is ....

1. Your basic salary at the commencement will be at the rate of
    £ 1300 per annum and will be payable ~~Monthly/Fortnightly~~. weekly
    You will be considered for increments at the end of each year
    of service under this Agreement.

2. You will be eligible for all benefits under the Company's
    Non-Contributory Retirement Assurance Plan. If you are not
    at present a member of this scheme, your date of joining will
    be the first day of the month of your appointment or transfer
    as the case may be.

3. The cost of travel for yourself, wife and children under 21
    years of age will be arranged and paid by the Company.

4. The Company will also arrange and pay for the transport of
    your furniture and personal effects within reason.

5. A Resettlement Allowance will be payable at the rate of £ 60
    per month for the first year/~~three months~~ and at £15   per
    month for the remainder of the time. This payment will
    commence on the date that your basic salary commences.

6. After taking up permanent duty in the A.C.T., an Operations
    Allowance will be paid for each day or shift actually served
    at the Station as under -

    Day Workers - 25% of basic day's pay - i.e. 25% of 1/5 of
    1/52 of Basic Annual
    Rate.

    Shift Workers - 50% of basic day's
    pay - i.e. 50% of 1/5 of
    1/52 of Basic Annual
    Rate.

    This extra payment is compensation for the slightly longer
    working hours at this Station and, in the case of Shift
    Workers, is also designed to replace the usual shift work
    allowance, as provided in Industrial Awards. This allowance
    also covers the disability of being on call in case of need.

    It is not payable during periods of Sick Leave or Annual Leave
    or Long Service Leave, nor on any day spent away from the
    Station, whether on duty or not.

    It should also be realised that although the Company does not
    recognise a liability to allow financial compensation for what
    may be claimed as excessive travelling time between the Station
    and Canberra (35 miles or so), the scale of the Operations Allow-
    ance has been raised to cover this disability as well as others.

7. The Normal Working Hours at the Station will be -

    Day Workers - a five day week, Monday to Friday of 38 working
    hours.
    Shift Workers - Rostered shifts will be arranged to average 21
    shifts per 28 days, to cover round the clock

requirements. Overtime will not be payable
for a twenty minutes hand-over period as this
is also covered in the 50% Operations Allowance.

8. Overtime Payments -

Overtime will payable as follows -

Mondays to Saturdays - For all time outside regular daily
hours or rostered shift - at rate and one-half basic hourly,
i.e. at 1.1/2 of 1/38 of 1/52 Annual Rate.

Sundays (Day Workers - all time at double rate.
(Shift Workers - all time outside normal rostered
shift hours at double rate basic
hourly, i.e. at: 2/1 of 1/38 x
1/52 of Annual Rate)

Gazetted Public Holidays -

Day Workers - For normal daily hours worked -

Double daily rate, that is an extra day's
pay in addition to the day paid as Public
Holiday pay.

For work outside normal daily hours -
Double basic hourly rate.

Shift Workers - For work outside normal rostered shift hours
Double basic hourly rate.

9. Annual Leave

Three weeks' Annual Leave is allowed to Day Workers for each
completed year of service. Shift workers will receive
additional leave at the rate of 1/10 of a week for each
Sunday worked on shift with a limit of 1 week for each 12
months period.

10. Home Leave

A married employee who takes up duty at the Data Acquisition
Facility after his initial transfer to Canberra and before
he has been able to make permanent housing arrangements will
be allowed to return once to his former place of residence
in Australia to arrange his family's transfer or he may
travel to Canberra and return for the purpose of arranging
housing there, if convenient before his official starting
date on the Contract.

11. Sick Leave

Sick leave entitlement accrues at the rate of three weeks per
annum on full pay for each year of service and is cumulative.

12. Travelling Allowances

The following allowances are payable to employees on short-
term duty away from their permanent station overnight:

		Daily Allowance
12.1	Within Australia	
	1) Capital Cities (and Carnarvon)	£4 5 0
	2) Country (including Woomera)	2 14 0
12.2	In U.S.A. (up to 6 months)	
	1) New York and Washington	US$ 23.00
	2) Elsewhere	US$ 23.00

13. It must be understood that the matter of housing for yourself and family is basically your own private problem. The Company is prepared to assist you with information and advice.

14. You are also offered a Formal Service Agreement, which, with this letter will constitute the Contract between yourself and the Company.

15. You must undertake to reimburse the Company for all expenditure incurred under Clauses 3, 4 and 5 and also the expenses involved in providing you with training overseas, if you do determine your employment except by mutual consent.

    If you accept this offer and are agreeable to all of the conditions, please return the copy, duly signed, as soon as possible.

Yours faithfully,

E.M.I. ELECTRONICS LIMITED

. . . .
Signed

. . . .
Date

Appendix 4 – A brief outline of how spacecraft radio signals were received at Orroral tracking station.

For practical reasons at each STDN tracking Station it was often necessary for the actual data receivers to be situated quite some distance from the receiving antenna.

This meant that there could be long lengths of cable, sometimes more than a kilometre, between the antenna and the receiver. Unfortunately, due to the laws of physics, over such a long distance cable will reduce (attenuate) the signals and degrade them before they reach the receiver.

So a method had to be used that overcame this cable degradation of the signals.

Right at the very heart of each receiving antenna was a very special and sensitive amplifier called a Low Noise Amplifier (LNA). This amplified the signal directly as it was received and passed it to another device at the antenna, called a converter.

The converter changed the signal to a frequency which was more suitable for sending through a cable. It also amplified the signal by an amount calculated to overcome the losses in the cable.

The signal was sent through the cable to the receiver area of the station, where it was further amplified and divided in a device called a 'multicoupler', so that it was available to all receivers on the station. This equipment configuration allowed the tracking station to connect any receiver to any signal from any antenna.

The result of this careful design and engineering was such that the received signal was applied to the receiver almost as if the receiver was connected directly to the antenna and there was nothing between them, even though they could have been separated by hundreds of metres, or more, of cable distance.

Appendix 5 - Equipment setup and local tracking schedules

How all of the equipment at any tracking station throughout the network was to be set up and used for each and every spacecraft was set out in a document called the Network Operations Support Plan (NOSP).

There was a NOSP for each spacecraft tracked by the network. Depending on the complexity of the spacecraft, this document may have ranged in size from about 30 pages, to many hundreds of pages. The NOSP contained all of the settings for all of the equipment at all of the tracking stations.

The NOSP for the Space Shuttle was two volumes of more than 300 pages each. Such a document was too unwieldy (and confusing) to be directly used in the operations area. The procedure adopted at Orroral was to simplify this information for use by the operations staff.

The station's operations support office, headed by the Senior Operations Supervisor, extracted all of the information relevant to the Orroral Tracking Station and produced a much smaller document called a 'Local Operations Procedure' or LOP. Multiple copies of this document were made available throughout the operations area, so that each operational area had access to the information.

Details gained by staff who had travelled to other NASA tracking stations outside of Australia, led me to believe that the Orroral LOP procedure was unique.

The LOP was then divided into further tabbed sections, which detailed the equipment configuration and procedures for each of the distinct operational sections of the station. It also included any additional information in the form of teletype messages or notes which were not incorporated into the NOSP document. This meant that staff had easy access to all of the information relevant to their section. Because the information was not cluttered with details about other stations or different equipment types, it also reduced the likelihood of errors.

Project Boards

The project board was a matrix board of 128 x 128 holes into which special coaxial patch cords could be inserted. The cords were tipped with blunt-pointed plugs that protruded through the panel, so that when the board was inserted into its rack, the ends would make contact with a terminal array on the backplane. This enabled all of the required inputs and outputs of the equipment to be connected together in the correct way for each of the different spacecraft.

The gross equipment connection was then made by push-button selections on the crossbar switching system panel. The crossbar switching system had the capability to connect all of the 16,384 hole positions on every project board, for every telemetry, command, and data handling system, and each piece of equipment in those systems throughout the station. It was very complex. The crossbar switching system was the equivalent of a telephone exchange for a moderately sized town.

The LOPs for all spacecraft supported at Orroral were of the same basic layout and structure. However, not all sections were needed for all of the spacecraft tracked by Orroral. Those sections that were not required were simply omitted.

The Space Shuttle LOP was one of the most complex ever used at Orroral. It contained about 190 pages.

Local Operations Procedures (LOPs)

The Space Shuttle LOP.

Orroral's Space Shuttle LOP contained the following sections:

- At the front of the LOP was a list of applicable messages.
- Section 1. General information about the spacecraft and flight. It also contained the pre-pass H-45 sequence. It was 22 pages long.
- Section 2 Was for the receiver section. It contained all of the receiver settings and tests. It was 5 pages long.
- Section 3 Detailed the magnetic tape recorder section and contained all of the configuration and speed settings for the data tape recorders, both for recording data during the tracking pass and playback of data after the tracking pass had finished. It was 16 pages long
- Section 4 Was the S-Band transmitter and ranging system. It was 19 pages long.
- Section 5 Computers. 642B computer systems and peripherals. 15 Pages in length.
- Section 6 Command. Spacecraft Command Encoders and Space Shuttle voice multiplexer was 4 Pages
- Section 7 Data Handling. PSK demodulators, MSFTP-2 decommuntators, DDPS word formatter, Frame synchronisers. Multiple configurations for both real-time data and post-pass playback of data. 37 Pages in length.
- Section 8 Air-Ground Voice. ComTech console and air-ground voice recorder configuration. Normal and special procedures as required. 13 Pages in length
- Section 9 Ground communications voice procedures. Procedures to be used for voice communication circuits was 9 Pages
- Section 10 Procedure for interface to Mission Control Centre (MCC Houston), Network Operations Control Centre (NOCC Goddard), and other stations as required, including teletype messaging was 8 pages
- Section 11 Recorded Data Disposition. Labelling, documenting and shipping requirements for tape recorded data. 3 pages in length.
- Section 12 Operations Supervisor unique requirements. 18 Pages in length.

To show how complex the equipment setup and procedures were for the Space Shuttle, some sections of that document are shown in this appendix.

Local Tracking Schedule

Another procedure that the author believed to be unique to Orroral, was the way that the tracking operations were scheduled locally. Comments made by visitors to the station led the author to believe that the way tracking operations were scheduled at Orroral was very different from other tracking stations throughout the world.

It seemed that most stations in the network used the rather lengthy (and often hard to read) schedule which was sent by teletype each day from Goddard Space Flight Centre, but with some comments locally added and other guidelines written in. At Orroral the night shift supervisor used the schedule received from GSFC to produce a much easier to read and clearer document which made the tracking operations less prone to error. By 1975, this schedule had been changed from a multipage format to a bar chart format. Two bar charts were made to cover each eight-hour shift period.

One chart covered tracking systems one and two, and the second covered tracking systems three, four and five. As well as labelling, colour coding was used to identify the system. On the link 1-2 schedule, link one was shown in red and link two in black. On the link 3-4-5 schedule, link 3 was blue, link 4 was green and link 5 was brown. Copies of the schedule were distributed to each operational area of the station. An example of a schedule for links 1 and 2 for one shift period is shown in chapter 8.

Extracts from the Space Shuttle LOP

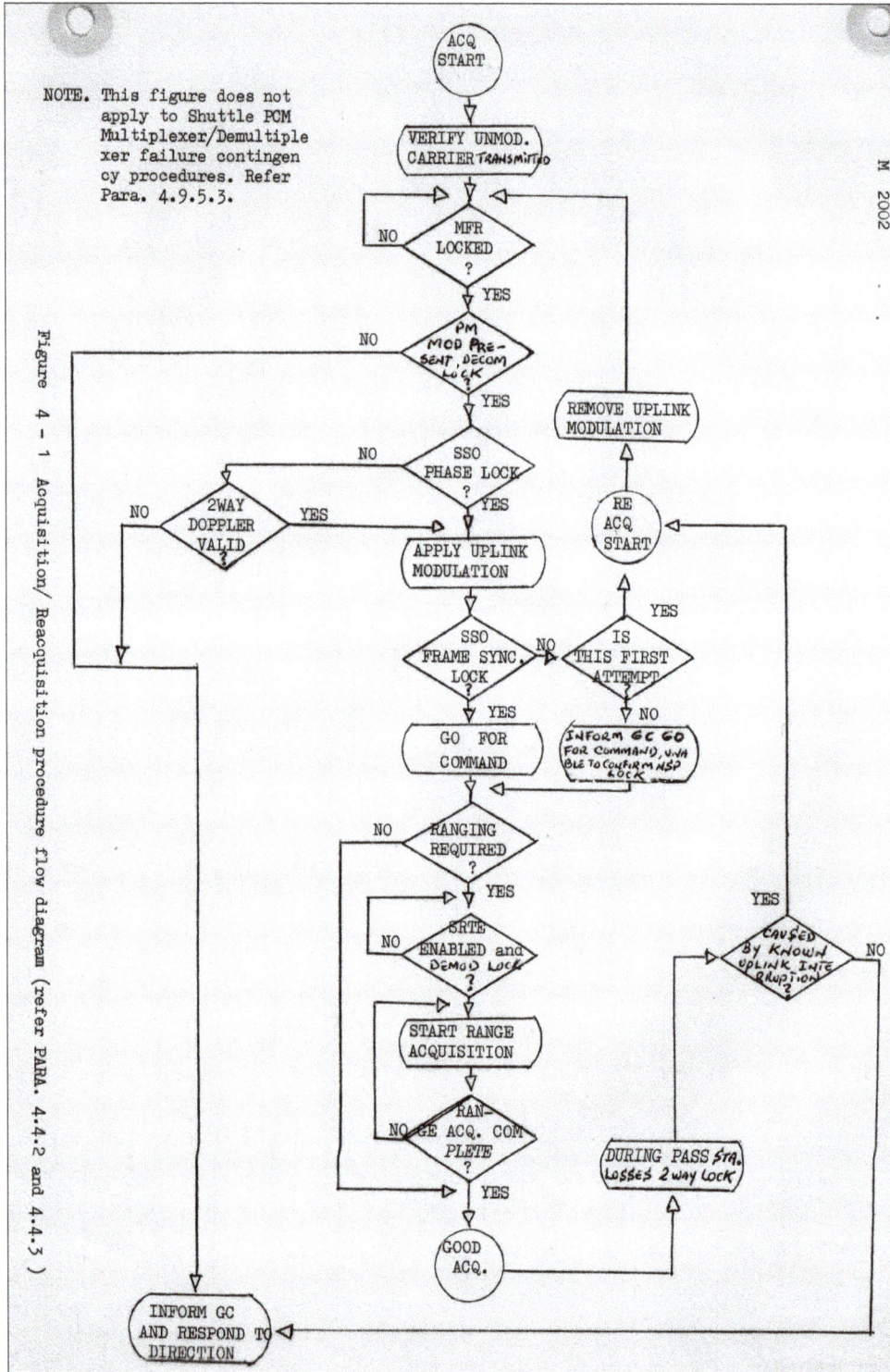

Figure 4 – 1 Acquisition/ Reacquisition procedure flow diagram (refer PARA. 4.4.2 and 4.4.3)

Table 7-6   Decom Program Format Select Versus SCM Downlink Modes

SCM Downlink Modes			Program								Data Type	SOFT (642B) HSP Advisory Message
			Simulator (Note 5)	Data Gen (Note 5)	MSFTP-3 (Note 5)		MSFTP-2 (Note 5)		MSFTP-2 (Note 5)			
OD DL	FM DL	DF DL	Format	Format	Format	PCM ID	Format	PCM ID	Format	PCM ID		
03,05			01	143	01	01	01	01			192-kb/sec OD real time	OD RT PB HBR (note 2) TCT OD RT TLM HBR
02,04			02	144	02	02	02	02			96-kb/sec OD real time	OD RT PB LBR (note 2) TCT OD RT TLM LBR
	02		01	173	10	10			03	10	192-kb/sec OD Dump Monitor (Forward)	OD MON FWD 192 kb/sec
	03			150	11/23	11					960-kb/sec OD Dump Monitor (Forward)	OD MON FWD 960 kb/sec
	04		03	147	07	07			02	07	128-kb/sec OD Dump Monitor (Forward)	OD MON FWD 128 kb/sec
	05			151	12/23	12					1024-kb/sec OD Dump Monitor (Forward)	OD MON FWD 1024 kb/sec
	12		06	153	14	14			07	14	192-kb/sec OD Dump Monitor (Reverse)	OD MON REV 192 kb/sec
	13			154	15/23	15					960-kb/sec OD Dump Monitor (Reverse)	OD MON REV 960 kb/sec
	14		07	155	13	13			06	13	128-kb/sec OD Dump Monitor (Reverse)	OD MON REV 128 kb/sec
	15			156	16/23	16					1024-kb/sec OD Dump Monitor (Reverse)	OD MON REV 1024 kb/sec
		M	03	147	17	17			07	17	128-kb/sec DFI Monitor	DFI Monitor
NA	NA	NA	04	145	05	05	05	05			72-kb/sec High-rate Command (note 3)	CMD HIST GMT TCT

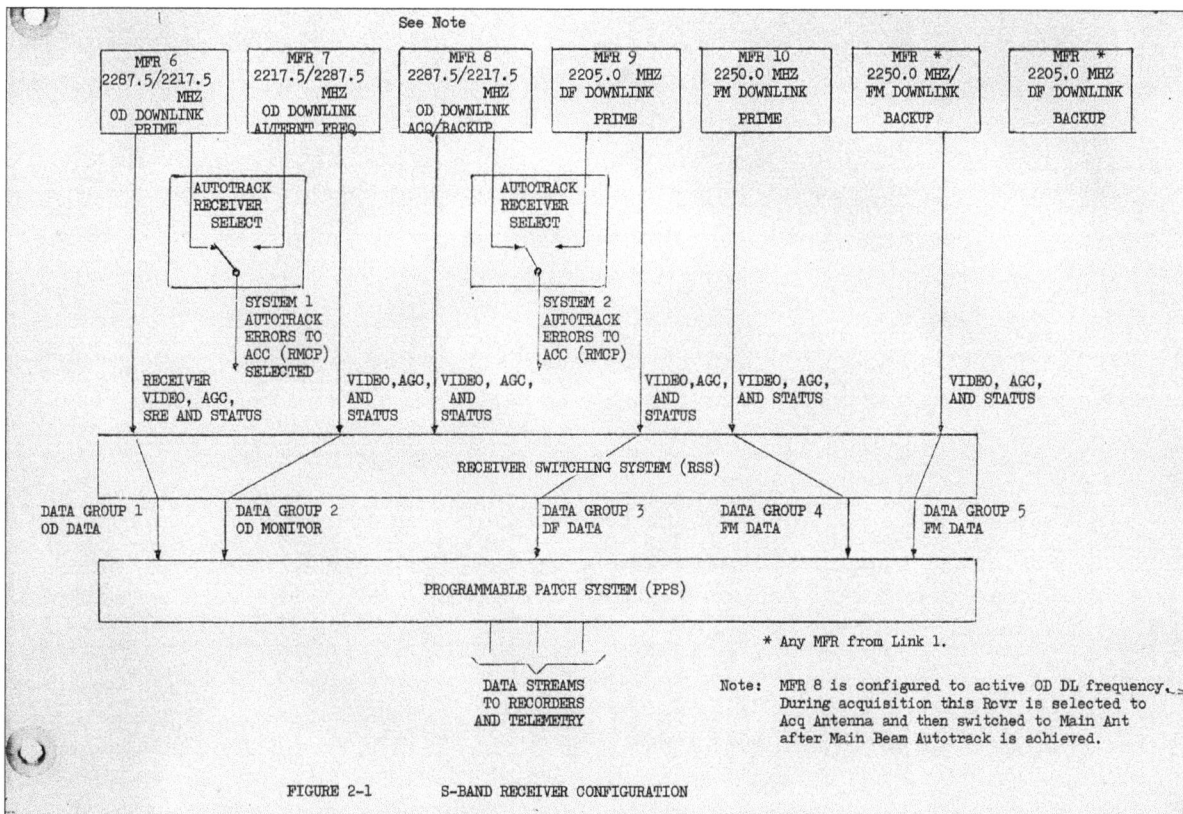

FIGURE 2-1   S-BAND RECEIVER CONFIGURATION

Appendix 6 - Transcript of Voice Communications During a H-45 (Horizon minus 45 Minute) Sequence at Orroral Valley Tracking Station.

The Story Behind the Tape Recordings.

In a previous chapter and in appendices 6 and 7 are transcripts of tape recordings of voice communications at the Orroral Valley Tracking Station which was made during the first orbital flight of the Space Shuttle and the preparation for that flight. I believe that this recording is unique and contains material that even NASA has not archived. Here is how the tape came to be.

After I had researched the book and made the draft, I was cleaning up some unrelated boxes of material. In one box I found an unlabelled two-hour cassette tape. When I played the tape, I discovered that it was a recording of communications at Orroral during the first flight of the Space Shuttle. I found that I had recorded six passes or tracks of the space shuttle at the Orroral Valley tracking station. The orbits recorded were:

Orbit 2	Orbit 16
Orbit 14	Orbit 17
Orbit 15	Orbit 18.

I have no memory of making the tape, or why I made it. At the time I am certain that I would not have been thinking of preserving something for posterity. I am sure I would have only made it for personal interest as something that I did. As to how I made the recording I cannot actually remember. However, the quality leads me to conclude that I did more than simply place the recorder in front of a loudspeaker. The recording quality could have only been achieved by somehow connecting the recorder directly to the communication monitoring circuits. I cannot recall exactly how I might have done this, but at that time I was quite adept at electronic innovation and adaptation. I know that NASA has archived the communications between the Houston Mission Control Centre (MCC) and the astronauts on the Space Shuttle. I am not sure they archived the communications between the tracking stations and the control centres in the United States. I believe that what is unique about this recording is the inclusion of the communications within the operations area of the Orroral Valley Tracking Station.

On looking back, I think I may be able to reconstruct how I made the recording. It has to do with the configuration of the Air/Ground communications console at the station and the fact that I was the supervisor for voice communications. This last point gave me access to the equipment and circuits. The Air/Ground communications console was provided with a 9-track tape recorder that recorded the following:

Track 1	On station communications
Track 2	Off station communications (Site coordination circuit)
Track 3	Air/Ground 1 circuit
Track 4	Air/Ground 2 circuit
Track 5	Air/Ground 1 uplink verification
Track 6	Air/Ground 2 uplink verification
Track 7	36 bit time code
Track 8	Air/Ground 1 uplink monitor
Track 9	Air/Ground 2 uplink monitor

The tape used was one inch wide and 3,600 feet long. The tape was run at a speed of 1 7/8 inches per second for about six hours and changed, or the period of Space Shuttle tracking for each group of passes at Orroral, whichever was appropriate. I believe that what I did was to tap into the lines that went to tracks 1, 2, and 3 on the recorder and fed them into a cassette recorder through a little adaptor of my own making. What happened to these 9-track tapes after the Space Shuttle flight I do not know, but I have never found any reference to them in any NASA records I have searched. I strongly suspect that by the time of writing this book they have never been reviewed or used and have now been lost forever. It is possible that my recording is the only such record that remains and is therefore of significant historical value. For this reason, I have digitised them and preserved them as files and on CDs.

The H-45 Interface Sequence

The following transcript is from a tape recording of three communications circuits recorded on a single-track recorder. Therefore, there are a few occasions where conversations overlapped and could not be properly deciphered. As near as possible, the printed sequence of events is shown with the appropriate part of the tape transcript to allow the reader to have some insight as to how the operation ran. The communications circuits recorded were:

1. The off-station communications between Orroral, the control centres in the United States, and other stations. (External circuit.)
2. The air to ground number one circuit between Orroral and the MCC Houston (Air/ground circuit.)
3. The internal on-station communications circuit within the Orroral Valley Tracking Station.

As far as possible this transcript follows the actual conversations on all of these circuits in correct time sequence.

The following pages are the transcript of the H-45 sequence, and in Appendix 7 you will find the transcript of the actual shuttle orbit communications.

Scanned pages of the operating procedure are interspersed with the actual transcript, so that you can follow the flow of activity.

Most call signs are self-explanatory, but where necessary definitions have been given.

- REC: is Orroral receiver technician,
- USB is Orroral unified S-Band technician,
- DH is Orroral data handling technician,
- Opsr is Orroral Operations Supervisor, Peter Uzzell.
- Except where noted as GCC the COMTECH is Philip Clark. (Author).
- RTC is Houston Real Time Command controller,
- DFE is Houston Data Flow Engineer.
- NST is Goddard Network Support Team.
- GCC is the communications centre operator at Orroral tracking station.

The H-45 (Horizon minus 45 minute) Sequence Transcript.

Local Operating Procedure

Table 1-2   STDN H-45 Interface

Time	Position	Sequence	Action
H-45	NOM/OPSR Com Mgr	1	Com Mgr confirm lines established between GSFC and station. Station confirm completion of HSD line test block transmission on all 56-kb/sec or 224-kb/sec data lines.
	NOM	2	NOM request Voice Control add station to NOCC interface CKT. NOM conduct voice and status check. Confirm receipt of SCM and ACQ data. Station systems loaded per: Computer (TLM) SSI No._____ PCM SSI No._____ Command SSI No._____ Track SSI No._____ Note. When SCM indicates PAYLOAD DUMP, the PAYLOAD DUMP RATES will replace the 960/1024KB SHUTTLE DUMP checkouts respectively.
	OPSR	3	Confirm 642B computer in pass phase (A-111) DLSM enabled, SSM logging initiated. SCE in pass mode, SCVM SAFE/OPERATE switch set to SAFE and the S-band carrier is up at mission power level and radiating into the dummy load, with modulation ON. Enable AMQ. Select data generator format ___. 642B computer destination code is 064. Notify NOM when completed.
H-42	NOM/OPSR	4	Stand by for NOCC interface.
	NST A-G/Com Tech	5	Perform keying and modulation checks on A-G 1 and A-G 2 S-band and UHF. Verify downlink tone levels on UHF A-G1 and A-G2.
	OPSR	6	Select DLSM parameter list and confirm constants. CAM (G,444).
	OPSR	7	Select parameter listing and confirm constants. CAM (G,111).
	OPSR/NST Data	8	On cue from NST data, enable MCC lines and SSM to MCC. Notify NST data that sequence 6 and 7 are completed.
	NST Data	9	Enable NOCC lines 1, 2, and 3.
	NST Data	10	Turn on Site Status MSG (SSM) to the NOCC.

Tape transcript

Tape Time (Min/sec)	Callsign	Transcript/ Communications on air/ground circuits.
13:06.	NST	Orroral comtech NST comtech air to ground airground 1
13:09	ORRORAL COMTECH	Orroral comtech airground 1.
13:11	NST	I read you 5 by, how me?
13:14	ORRORAL COMTECH	Roger you're 5 by.
13:15	NST	OK ready for keying & modulation?
13:19	ORRORAL COMTECH	That's affirmative.
13:20	NST	Alright. Garble. (Tones only, six keys)
13:37	ORRORAL COMTECH	NST air/ground keying was 100%.
13:42	NST	Roger, modulation check. (With tones) NST testing on air to ground 1, One, two, three four, five, four, three, two, one, test off
14:05	ORRORAL COMTECH	NST air/ground Orroral comtech, keying was 100%, voice & modulation were go.

Communications on external circuit.

14:05	NST?	Orroral what is your status for your upcoming H-45
14:16	Opsr	Our status is green,
14:16	NST	Roger, I copy, stand by for air ground 2
14:18	ORRORAL COMTECH	Roger.
14:18	NST	Roger sir I would like to confirm you have completed sequence 1 at this time.
14:21	Opsr	That is affirmed.
14:21	NST	Voice NST air to ground on air to ground checkout.
14:22	NST	I would like to confirm you are in receipt of acq data and SCM.
14:26	Opsr	Affirm.
14:28	NST	and confirm that your computer systems are loaded with SSIs 169 telemetry, 107 PCM, 82 command, and 133 track
14:34	NST	Voice NST air to ground on air to ground checkout
14:41	Opsr	Copy, that's affirmed on all.
14:44	NST	Ok sir, why don't you proceed with sequence 3 and advise me when it is complete.

14:49	Opsr	Sequence 3 is complete.
14:52	NST	And confirm destination code 064.
14:55	Opsr	We have a 064.
14:53	VC	Voice
14:55	NST	Roger voice can I get Orroral's air to ground 2 at this time.
14:56	VC	Roger.
14:57	NST	OK standby for NST data.
14:58	VC	You have Orroral's air to ground 2.
14:59	ND	Orroral NST data.
15:01	Opsr	Orroral
15:02	NST	Orroral comtech.
15:02	NST	Yes sir are you ready to (garble)
15:05	Opsr	That's affirmed.
15:07	NST	OK Orroral go ahead and enable the MCC lines and SSMs please

On-station communication

15:21	M&O	Computers M&O
15:23	COMP	Computers
15:24	M&O	Roger have you started logging yet?
15:26	COMP	That's negative, I'll do that now.
15:27	M&O	Righto, Thankyou

Communications on external circuit

15:38	Opsr	NST data Orroral.
15:40.	NST	NST data
15:41	Opsr	Sequence 8 is complete.
15:43	NST	Roger Orroral, we see your SSMs. Standby for NST SCE & NST telemetry.
15:49	NST	Orroral NST SCE.
15:50	Opsr	Orroral.
15:53	NST	OK would you re-initialise your SCE at this time.
15:56	Opsr	Wilco.
16:03	NST	Orroral NST telemetry. Proceed with sequences 12 & 13 and advise telemetry when ready.
16:08	Opsr.	Roger

On-station communication

16:10	M&O	Receivers M&O
16:11	REC	Receivers
16:12	M&O	Roger do you have all links set up?
16:14	REC	All links are set up minus 90 dbm.
16:18	M&O	Roger copy. Data handling M&O.
16:19	DH	Data handling.
16:19	M&O	are you locked on all 3 decoms?
16:21	DH	That's affirmative.
16:25	M&O	Roger.
16:28	M&O	Data handling M&O remove inhibits from the DFI decom only please.
16:33	DH	Inhibits removed.

Communications on external circuit

16:41	NST	Roger Orroral, we see your SCE re-initialised would you configure for low data rate uplink
16:46	Opsr.	Copy low data rate

On-station communication

16:51	M&O	SCE M&O configure for low data rate uplink.
16:51	SCE	Roger.

Communications on external circuit

16:49	NST	Orroral NST telemetry we see good DFI. That completes sequence 14, proceed with sequence 15 & 16 copy when you are ready.
16:59	Opsr	Roger

On-station communication

17:00	M&O	Data handling M&O, inhibit your DFI decom please.
17:04	DH	Inhibit is on.

Communications on external circuit

17:07	NST	Roger Orroral we see low data rate uplink, Err, We do not see DMS lock.

On-station communication

17:19	Comtech	M&O comtech
17:21	M&O	M&O

17:21	Comtech	Yes. Can we have the uplink back please you took it away in the middle of the air/ground checks.
17:27	M&O	Stand by.
17:31	Comtech	High data rate.
17:32	M&O	NST has requested low data rate. We will be going back to high data rate shortly.
17:39	Comtech	Er, OK, well NST has asked me for high data rate on the air/ground checks. Apparently, they are not synchronised at their end. However, it did go in the middle of our check and made a mess of it.
17:54	M&O	OK
17:59	M&O.	Data handling M&O
17:59	DH	Data handling
18:00	M&O	Can you re- configure the DMS for low data rate.
18:06	DH	It was configured for high data rate. It's now configured for low data rate. Do you want it back to high data rate or stay in low?
18:12	M&O	Is it locked up?
18:14	DH	It's locked er, I've got bit sync lock but no frame sync lock.

Communications on external circuit

18:20	NST	Orroral NST SCE
18:23	Opsr	Orroral.
18:23	NST?	Ok you have a problem with DMS lock
18:27	Opsr	OK we're not showing lock at the time, there was a conflict of interest between yourselves and the air/ground people. Stand by please.
18:39	Opsr	Er, We'd like to re-initialise the SCE MUX if we may please.
18:46	NST	Roger.

On-station communication

18:49	M&O	SCE M&O
18:50	SCE	SCE
18:51	M&O	Roger, re-initialise your SCE MUX.
19:17	SCE	SCE MUX is re-initialised.

Communications on external circuit

19:17	NST	OK Orroral. We see lock on low data rate. Will you perform SCE loop test?
19:37	Opsr	NST SCE Orroral.
19:39	NST	NST SCE

19:40	Opsr	Loop is good, PEP is good, all the flags are zeros.
19:44	NST	Roger, reconfigure for high data rate uplink.
19:48	Opsr	Wilco.

On-station communication

19:49	M&O	SCE M&O, reconfigure for high data rate uplink.
19:51	SCE	Wilco.
19:52	M&O.	Data handling M&O
19:54	DH	Data handling.
19:55	M&O	Reconfigure for high data rate.

Local Operating Procedure

Table 1-2   STDN H-45 Interface (cont)

Time	Position	Sequence	Action
H-42 (cont)	NST SCE/OPSR	11	a. Select LDR or HDR as directed by NST SCE. b. Set SCVM SAFE/OPERATE switch to OPERATE. c. NST SCE transmit Uplink Test Command (UTC) and verify station val cap. d. Set SCVM SAFE/OPERATE switch to SAFE.
	OPSR/NST TLM	12	Configure the test setup to simulate the DFI monitor signal. Inform NST TLM when configured.
H-40	NST TLM/OPSR	13	On cue from NST TLM enable the DFI monitor test signal into the RF loop at -90 dBm. Decom inhibit OFF. Confirm decom lock on HSP.
	NST TLM	14	Verify on DTV display. PCM lock AGC level equals -90 dBm; frame/sec equals 100.
	NST TLM/OPSR	15	On cue from NST TLM disable the DFI monitor test signal.
	NST TLM/OPSR	16	a. Verify that real-time OD HBR signal inputs are modulating signal generators from PCM simulator. On cue from NST TLM, enable the test signal into the RF loop at -90 dBm. Decom inhibits OFF. b. Configure the data generator input of OD dump 960 kb/sec (reverse) to FM signal generator 2250 MHz at -90 dBm. Inform NST TLM when configured. Decom inhibits OFF. c. Confirm decom lock via HSP.

Table 1-2   STDN H-45 Interface (cont)

Time	Position	Sequence	Action
H-40 (cont)	NST Data/TLM	17	a. Verify on DTV display. b. PM rec lock _____; AGC level _____. PCM lock _____; frames/sec equals 100. c. Dump PCM lock ____; AGC level ____; Dump time ____. FMT ID _____; frames/sec equals 500. d. NST Data turn on TLM data to NOCC.
	NST TLM	18	Trap a frame. Verify contents on video display. Check SC and block time.
	NST TLM/OPSR	19	Remove test signals from RF loop and verify decoms inhibits are ON.
H-35	NST Data/OPSR	20	NST Data disable NOCC lines. OPSR disable MCC lines. Enter destination code 160.
H-30	NOM/OPSR	21	NOCC interface complete, stand by for MCC interface. (While NOCC/MCC is configuring SITE COORD and confirming interface status, the station should configure for sequence 25 and notify DFE when ready.)
	NOM/GC	22	Notify GC that station is ready to start interface. Either confirm go-ahead with MCC interface, or define limitation.
	NOM	23	Notify Voice Control to configure station to SITE COORD.
	Com Control Com Mgr	24	Confirm station data circuits configured to MCC.

Tape transcript

Tape Time (Min/sec)	Callsign	Transcript/ Communications on air/ground circuits.
19:57	DH	Reconfigured.
20:04.	DH	We have lock
20:06	SCE	SCE is reconfigured.
20:06	M&O	Roger.
20:07	M&O	Data handling M&O, remove your inhibits on the OD & FM dumps please.
20:15	DH	Inhibits removed.
20:16	M&O	Roger.

Communications on external circuit

20:22	Opsr	NST SCE Orroral,
20:24	NST	NST SCE.
20:25	Opsr	We are configured in high data rate.
20:28	NST	OK you can perform SCE loop test.
20:35	Opsr	NST telemetry Orroral.
20:38	NST	Telemetry.
20:39	Opsr	16 is complete.
20:42	NST	Roger, stand by one and we will verify.
20:43	Opsr	NST SCE Orroral.
20:45	NST	SCE
20:46	Opsr	Loop test on high data rate, loop is good, PEP is good, flags are zeros.
20:51	NST	Roger, safe the SCVM.
20:55	Opsr	SCVM is safe
20:58	NST	Roger, we see it.
21:10	NST	Orroral your data looks good for telemetry. That completes sequences 18 er 17 &18. Proceed with sequence 19 and stand by for data 20.
21:21	Opsr	Wilco

On-station communication

21:21	Comtech	M&O comtech
21:23	M&O	M&O.

| 21:24 | Comtech | Roger. We have completed air/ground tests, we have a go on the air/ground interface. |
| 21:29 | M&O | Roger, copy |

Communications on external circuit

21:29	NST	Orroral NST data.
21:30	Opsr	Orroral
21:32	NST	Roger. Disable your MCC lines. Disable NOCC lines enter & verify destination code 160
21:39	Opsr	Roger

On-station communication

21:30	M&O	Data handling M&O.
21:32	DH	Data handling.
21:33	M&O	Inhibit all decoms.
21:36	DH	Inhibited.
21:48	M&O	Data handling M&O
21:49	DH	Data handling.
21:50	M&O	Configure SOU2 for format 151 please.
21:54	DH	Stand by.
21:58	USB	M&O USB
22:00	M&O	M&O
22:01	USB	Can I take carrier down?
22:04	M&O	That's affirmative.
22:05	USB	Carrier is down, modulation is off.
22:07	M&O	Roger.
22:09	DH	SOU configured.
22:10	M&O	Thank you
22:11	USB	M&O USB.
22:12	M&O	M&O.
22:14	USB	Er, give the boys up here a good turn around.
22:19	M&O	(garble) Timmy. Er, sometime later we'll bear it in mind.
22:25	USB	Roger thank you.
22:35.	M&O	Data handling M&O
22:37	DH	Data handling.
22:37	M&O	Roger. Is your MSFTP-3 locked up now?

Local Operating Procedure

Table 1-2   STDN H-45 Interface (cont)

Time	Position	Sequence	Action
H-30 (cont)	OPSR/DFE	25	Verify station is configured for the MCC interface: a. Destination code 160_8. b. Data generator input of OD 192-kb/sec TLM to PM signal generator per SCM designated frequency at -90 dBm. c. Data generator input of OD dump 1024 kb/sec (reverse) to FM signal generator 2250 MHz at -90 dBm. d. PM receiver/decom locked with inhibits OFF. e. OD dump/monitor decom locked with inhibits OFF. f. DFI/FM receiver to a monitor decom with inhibits ON.
H-29	DFE/OPSR	26	DFE turn TLM ON (enable MCC lines) and turn SSM ON. a. Verify correct indication of station configuration. b. Signal strength equals -90 dBm. OD RT frame/sec equals 100 (see sequence 29). FM dump frame/sec equals 800
	DFE/OPSR	27	Verify station reports agree with MCC indication.
	OPSR	28	OPSR reconfigure data generator to the 128-kb/sec DFI format input to the PSK simulator and an FM signal generator for 2205 MHz at -90 dBm. Decom inhibits OFF.
	OPSR/DFE	29	a. Repeat steps 26a and b for the DFI FM link (all telemetry I/F sequence complete by H-15). b. DFI mon frame/sec equals 100.

Tape transcript

Tape Time (Min/sec)	Callsign	Transcript/ Communications on air/ground circuits.
22:40	DH.	That's affirmative

Communications on external circuit

22:41	Opsr	NST data Orroral.
22:43	NST	NST data.
22:45	Opsr	MCC lines are off. Destination code is a 160.
22:40	NST	Roger Orroral, stand by for data Com.
22:53	Goddard Ops.	Orroral Goddard Ops
22:54	Opsr	Orroral.
22:55	Goddard Ops	Alright sir we'll give you a go on you NOCC interface. Stand by and we'll put you on the site coord for your MCC interface
23:01	Opsr	Roger
23:07	Goddard Ops	Voice control, Goddard Ops NOCC interface.
23:11	VC	Voice.
23:11	Goddard Ops	Put Orroral to site coord please.
23:13	VC	Roger.

On-station communication

24:17	USB	M&O USB
24:19	M&O	M&O.
24:20	USB	On sequence 35 we show dummy load. Do you require dummy load or do you want antenna for that one?
24:26	M&O	Stand by.

Communications on air/ground circuits

24:30	HOUSTON COMTECH	Orroral Valley comtech Houston air to ground one.
24:33	ORRORAL COMTECH	Orroral comtech air/ground 1.
24:35	HOUSTON COMTECH	You're 5 by. Meet me on air to ground 2.
24:28	ORRORAL COMTECH	Roger.

Communications on external circuit

24:33	RTC	Orroral RTC.

24:34	Opsr	Orroral.
24:36	RTC	Configure for a H-30 interface
24:44	Opsr	Wilco.

On-station communication

24:50	M&O	USB M&O
24:51	USB	USB.
24:52	M&O	Roger are you configured for sequence 35?
24:55	USB	That's affirmed, all mode, carrier up.
24:59	M&O	Roger
25:00	USB	Roger, we are configured.

Communications on external circuit

25:01	Opsr	RTC Orroral.
25:04	RTC	RTC.

On-station communication

25:03	USB	We are to operate.
25:04	M&O	Roger that.

Communications on external circuit

25:06	Opsr	We are configured per sequence 35.
25:10	RTC	Roger.

Communications on air/ground circuits

25:14	HOUSTON COMTECH	Orroral Valley Houston you've got the teleprinter coming at you now.
25:19	ORRORAL COMTECH	Roger.

On-station communication

25:31	Comtech	GCC comtech on link 2
25:33	GCC	GCC

Communications on external circuit

25:34	RTC	Orroral RTC.
25:35	Opsr	Orroral.

On-station communication

25:34	Comtech	Roger. Could you confirm the frequency and level of that signal?

Communications on external circuit

| 25:37 | RTC | We have a good interface, you may safe your SCVM. |
| 25:41 | Opsr | Copy. |

On-station communication

25:47	USB	SCVM safe.
25:51	USB	M&O USB can I (garble).
25:43	M&O	Stand by one.

Communications on air/ground circuits

25:51	GCC	Break, break,
25:57	GCC	Houston comtech Orroral comtech. We have a frequency of 2056 a level of negative 12.
26:06	HOUSTON COMTECH	Negative 12 at the comtech console?
26:10	HOUSTON COMTECH	What have you got going into your DMS?
26:12	ORRORAL COMTECH	Roger, standby we are just setting that now.
26:15	HOUSTON COMTECH	OK

On-station communication

26:04	M&O	USB M&O loop (garble)
26:08	USB	Roger
26:44	Comtech	GCC comtech link 2.
26:49	GCC.	GCC
26:49	Comtech	Roger are you letting that tone come through to the console. I can't hear it.
26:52	GCC	No we broke it.
26:55	Comtech	OK, I had better ask him to put it back I didn't hear him break it. I hadn't finished setting the level here.
27:01	GCC	Well I told him to break when I read the frequency and the level of it.

Communications on air/ground circuits

27:06	ORRORAL COMTECH	Houston comtech Orroral comtech.
27:09	HOUSTON COMTECH	Go ahead.
27:09	ORRORAL COMTECH	Roger, could you put the tone back? We had not completed setting the level when it was taken off.
27:15	HOUSTON COMTECH	OK, you got it again.
27:17	ORRORAL COMTECH	Roger thank you. Stand by please.

27:19	HOUSTON COMTECH	And I do need the verification level also.
27:23	ORRORAL COMTECH	Roger that. Stand by.
28:04	ORRORAL COMTECH	Houston comtech Orroral comtech
28:06	HOUSTON COMTECH	Go ahead.
28:07	ORRORAL COMTECH	The level at the DMS is neg 22, the level at the verification is neg 20.

Local Operating Procedure

Table 1-2 STDN H-45 Interface (cont)

Time	Position	Sequence	Action
H-29 (cont)	DFE	30	Turn TLM OFF (disable MCC lines).
	OPSR	31	a. All decoms inhibits ON. b. 642B to postpass phase CAM (A,222). Print DLSM and verify contents. c. 642B to pass phase CAM (A,111) enter DLSM parameters, notify DFE that computer is in pass phase.
	DFE	32	Turn TLM ON (enable MCC lines).
	DFE	33	Enable SSM.
	MCC/Station Com Tech	34	a. Houston Com Tech contact station Com Tech on A-G 1 and start A-G remoting test. First circuit tested will be A-G 2. (JSC VAL Test No. 4201 and 4208). b. Reconfigure the DMS for AOS.
	RTC/OPSR	35	RTC request OPSR configure for the command interface. a. S-band RF carrier ON. Radiate into dummy load. b. Modulation ON. c. SCVM to OPERATE.
	RTC/OPSR	36	RTC verify station configuration via SSM.
	RTC/OPSR	37	Verify SCE parameter list.
	RTC/OPSR	38	RTC execute test command and verify Val cap.
	RTC/OPSR	39	RTC give station GO/NO GO on command interface.
	RTC/OPSR	40	RTC request OPSR SAFE the SCVM.
H-20	DFE	41	Reconfigure TLM system for AOS.

Tape transcript

Tape Time (Min/sec)	Callsign	Transcript/ Communications on air/ground circuits.
		Communications on external circuit
28:14	Opsr.	That is affirmed
28:16	DFE	Madrid DFE OK Let me turn on your lines.
		Communications on air/ground circuits
28:20	HOUSTON COMTECH	Ok we'll go ahead and check the S-Band voice on air to ground 1 and you can stay configured with the teleprinter on air/ground 2
28:30	ORRORAL COMTECH	Roger.
28:33	HOUSTON COMTECH	OK let me send you a mark on air to ground 1.
28:37	ORRORAL COMTECH	Roger.
		Communications on external circuit
28:38	DFE	Madrid your lines are on.
28:42	Opsr	Orroral sees lines on.
28:44	DFE	Copy that stand by just a moment.
		Communications on air/ground circuits
28:50	GCC	Break, break,
28:53	GCC	Houston comtech Orroral comtech we have a frequency of 2524 level negative 8.
29:01	HOUSTON COMTECH	Here is a space.
		Communications on external circuit
28:59	DFE	Orroral I would like you to proceed with sequence 25 and give me a destination code of 160 please
29:04	Opsr	You have a 160.
29:07	DFE	And I am ready for your data generator please.
29:14	Opsr	3 data streams OK?
29:17	DFE	That would be just great.
		On-station communication
29:09	M&O	Data handling M&O remove the inhibits on three decoms please.
29:13	DH	Copy

Communications on air/ground circuits

29:14	GCC	Break, break,
29:17	GCC	Houston comtech Orroral comtech we have a frequency of 2474 level negative 7.5.
29:27	HOUSTON COMTECH	OK let me send you a series of keys.

Communications on external circuit

29:42	DFE	Orroral DFE. Your OD is green. We'll give you a go for support I'll leave you lines on and have RTC come to you

Communications on air/ground circuits

29:42	HOUSTON COMTECH	Orroral valley Houston.
29:44	ORRORAL COMTECH	Orroral
29:46	ORRORAL COMTECH	Keying was 100% on air/ground 1
29:49	HOUSTON COMTECH	OK we might as well check out his air to ground 2 line also if you are ready I'll send you a mark on that.

Communications on external circuit

30:01	OPSR.	We have done the initial interface with RTC
30:06	DFE	OK copy that then lines are going off and like to print out your DLSM.

Communications on air/ground circuits

30:10	GCC	Break, break,
30:13	GCC	Houston comtech Orroral comtech we have a frequency of 2524 level negative 11.
30:22	HOUSTON COMTECH	OK here's a space.
30:32	GCC	Break, break,
30:35	GCC	Houston comtech Orroral comtech we have a frequency of 2474 level negative 12.
30:43	HOUSTON COMTECH	OK let me send you a series of keys here on air to ground 2
30:50	ORRORAL COMTECH	Roger.
31:02	ORRORAL COMTECH	Houston comtech Orroral comtech keying 100% on air/ground 2.
31:07	HOUSTON COMTECH	OK now I guess I had better give you keying and modulation check if you are ready.
31:12	ORRORAL COMTECH	Roger, are you going to multiaccess both?
31:15	HOUSTON COMTECH.	Yes, I am
31:24	ORRORAL COMTECH	Houston comtech Orroral comtech, go ahead.

On-station communication

31:25	M&O	Comtech M&O
31:29	Comtech.	Comtech
31:30	M&O	Do you still need carrier up?
31:32	Comtech	That's affirmative, we're checking.

Communications on air/ground circuits

31:30	HOUSTON COMTECH	(with tones) Houston comtech testing one, two, three, four, five, four, three, two, one, test out.
31:46	ORRORAL COMTECH	Houston comtech Orroral comtech keying was 100% on air/ground 1 & air/ground2, voice is go on both air/grounds
31:54	HOUSTON COMTECH	OK now if you would give me downlink tones
31:56	ORRORAL COMTECH	Just stand by please.

On-station communication

31:59	Comtech	M&O comtech we have finished with the uplink and data handling could you give me sense switch 4 on SOU 2, er SOU 1
32:06	M&O	USB M&O.
32:07	USB	USB
32:07	M&O	Terminate the carrier.
32:10	USB	Roger, carrier is down, modulation is off.

Communications on air/ground circuits

32:16	HOUSTON COMTECH	Neg 12.5 on both circuits.
32:21	ORRORAL COMTECH.	Roger
32:22	HOUSTON COMTECH	OK looks good and we'll give you a go and configure for teleprinter on your air to ground 2, that is configuration Lima.
32:30	ORRORAL COMTECH	Roger, configuration Lima.
32:33	HOUSTON COMTECH	Roger that, OK thank you much.
32:35	ORRORAL COMTECH	Roger.

On-station communication

32:07	M&O	USB M&O
32:08	USB	
32:09	M&O	Terminate the carrier please
32:10	USB	Roger carrier is down modulation is off

32:16	USB	M&O (garble)
32:41	Comtech	M&O comtech.
32:46	M&O	M&O.
32:48	Comtech	Roger, we have a go on our Houston interface, we have changed the configuration, we'll be operating air/ground configuration Lima for teletype.
33:00	Opsr	Also, in the sleep configuration, is that right?
33:03	Comtech	I believe so.
33:04	Opsr	OK

Local Operating Procedure

Table 1-2   STDN H-45 Interface (cont)

Time	Position	Sequence	Action
H-15	Houston Track/ RTC/OPSR	42	Request the OPSR to configure for the track command I/F. a. Antenna positioned to the collimation tower. b. S-band RF carrier ON. Radiate into the antenna. c. Acquire two-way lock. d. Select autotrack. e. Modulation ON. f. Acquire Range.
	Houston Track/ OPSR	43	Houston Track verify station is configured for track I/F. Turn ON tracking data.
	Houston Track/ OPSR	44	Houston Track give OPSR GO/NO GO on the track I/F.
H-10	RTC/OPSR S-band UHF A-G	45	a. S-band RF carrier ON. Radiate into antenna with antenna positioned to the collimation tower. b. Modulation ON. c. SCVM SAFE/OPERATE switch to OPERATE. d. RTC execute uplink test command and verify station Val cap. e. Upon completion of the command and track interface checks, OPSR turn modulation and drive OFF and configure for support. f. All antenna's to initial point (IP) in AUX/IPT, program or manual mode, and confirm with OPSR.
	UHF A-G	46	Slave antenna to best source.
H-8	S-band	47	CAUTION Radiation restrictions are in effect. Do not radiate through the antenna until H-2 minutes, or as specified in the SCM.

Tape transcript

Tape Time (Min/sec)	Callsign	Transcript/ Communications on air/ground circuits.
33:06	Opsr	What, so you inhibit the QUINDAR and....
33:06	Comtech	I inhibit the QUINDAR on 1 and go with teletype on the other.
33:12	Opsr	OK
33:13	Comtech	It is a little unusual I guess, I think I had better go back and query him and make sure.
33:17	Opsr.	That's affirmed

Communications on air/ground circuits

33:25	ORRORAL COMTECH	Houston comtech Orroral comtech site coord
33:29	HOUSTON COMTECH	Houston comtech
33:30	ORRORAL COMTECH	Roger with this configuration Lima that we have just been advised, in view of the sleep configuration on the SCM I take it that we will inhibit the QUINDAR on air/ground 1 and go with the normal teletype configuration on air/ground 2. Is this correct?
33:45	HOUSTON COMTECH	That's correct.
33:46	ORRORAL COMTECH	Roger, understood.
33:47	HOUSTON COMTECH	Thank you Orroral.

On-station communication

34:00	M&O	Recorders M&O
34:01	RECORD	Recorders.
34:02	M&O	Roger could you normalise your DSS532 recorder please, and in a couple of minutes we'll be ready for cal.
34:10	REC	OK, give me a mark but we're ready anyway.
34:12	M&O	Roger.
34:20	RECORD	And 532 is back to normal configuration.
34:23	M&O	Roger, thank you.
34:20	Comtech	M&O comtech, airground 1 keying is inhibited at this time.
34:34	M&O	Copy thank you.

Communications on external circuit

34:57	Opsr	DFE Orroral

34:59	DFE	Go ahead Orroral.
35:00	Opsr	We'd like to put the signals around the loop to do a pre-pass calibration at this time. Is that OK?
35:08	DFE	Stand by just a moment.
35:12	DFE	Say that again what you want to do Orroral.
35:14	Opsr	Just put signals around our RF receive loop for pre-pass calibration.
35:19	DFE	That would be fine. Just be sure that I don't get any data generator telemetry on line.
35:24	Opsr	Wilco.

On-station communication

35:24	M&O	Data handling M&O.
35:24	DH	Data handling
35:25	M&O	Confirm all decoms are inhibited.
35:28	DH	That's confirmed.

Communications on external circuit

35:28	DFE	In fact, Orroral, just to sure that I don't get it I'm going to go ahead and turn your lines off, I'll re-enable them at H-5.
35:35	Opsr.	OK

On-station communication

35:31	9M	M&O 9 metre.
35:37	9M	M&O 9 metre.
35:38	M&O	M&O
35:39	9M	Which col tower do you want us to put this onto?
35:32	M&O	9 metre col tower.
35:34	9M	Roger.
35:46	M&O	And go to the col tower, disable both axis' and select autotrack.
35:53	9M	Roger, will do.
36:04	M&O	Receivers M&O.
36:05	REC	Receivers.
36:06	M&O	Are you ready for cals?
36:08	REC	Yes, we'll be starting on the minute.
36:12	Record	Receivers recorders.

36:15	REC	Receivers.
36:17	Record	We have our tape rolling.
36:39	9M	9 metre M&O, er 9 metre
36:47	M&O	9 meter M&O go ahead.
36:48	9M	We are on the 9 metre coll tower with axis' disabled and in autotrack.
36:55	M&O	Roger thank you.
37:12	M&O	Receivers M&O.
37:15	REC	Receivers.
37:15	M&O	As soon as you finish cals kill the RF loop please.
37:31	REC	That ends calibrations recorders.
37:34	Record	Roger, copy.
37:35	Record	And recorders back to flight speed.

Communications on external circuit

37:39	Track	Orroral track.
37:40	Opsr	Orroral
37:42	Track	Are you configured for sequence 42 of the H45 count?
37:46	Opsr	Stand by, we just have to lock up the ranging.

On-station communication

37:46	M&O	USB M&O
37:50	USB	USB
37:51	M&O	Can you set up for a zero-delay turnaround?
37:54	USB	Roger, standby
38:22	USB	M&O USB we are configured.

Communications on external circuit

38:27	Opsr	Track Orroral.
38:28	Track	Track go ahead.
38:29	Opsr	We are configured per sequence 42.
38:32	Track	Roger you can put your data on line whenever you are ready.

On-station communication

| 38:36 | M&O | 9 metre M&O put data on line. |

Communications on external circuit

38:40	Opsr	Data is on line.
38:42	Track	Copy

On-station communication

38:41	M&O	GCC M&O.
38:44	GCC	GCC
38:45	M&O	Confirm data leaving please, tracking data.
38:51	GCC	Roger, it's going now.
40:32	M&O	SCE M&O.
40:34	SCE.	SCE
40:35	M&O	Roger, we should come up shortly on a check of the H-10 sequence.

Communications on external circuit

40:38	Track	Orroral Houston track.
40:40	Opsr	Orroral.
40:42	Track	Your data looks good you can take it off line.
40:46	Opsr	Wilco

On-station communication

40:49	9M	Low speed data terminated.
40:55	USB	M&O USB can I terminate now?
41:01	M&O	Yes, you can probably take carrier down, no leave carrier up, just terminate ranging.
41:06	USB	Roger.
41:10	9M	M&O 9 metre servo.
41:12	M&O	M&O
41:12	9M	Do you wish us to go to zenith at this time?
41:16	M&O	That's affirmative.
41:18	9M	Roger.

Communications on external circuit

41:23	RTC	Configure for sequence 45 please,
41:26	Opsr	Wilco.

On-station communication

41:27	USB	SCVM to operate
41:29	M&O.	Carrier up, modulation on
41:31	USB	That's affirmed, we left it up.
41:33	M&O	Roger

Communications on external circuit

41:34	Opsr	RTC Orroral.
41:37	RTC	RTC.
41:37	Opsr	We're configured.
42:04	RTC	Orroral RTC.
42:05	Opsr	Orroral.
42:07	RTC	We have good interface test, you may drop modulation.
42:13	Opsr	Wilco.

On-station communication

42:13	M&O	USB Modulation off please
42:15	USB	Roger, carrier is down, modulation is off.
42:23	M&O	9-meter M&O.
42:24	9M.	9 meter
42:25	M&O	You can go to intercept angles now.
42:27.	9M	Intercept angles, roger
42:31	M&O	Data handling remove decom inhibits.
42:32	DH	Inhibits off.
42:35	USB.	USB carrier is up
42:37	M&O	Roger that,
42:37	9M	Low speed data on line
42:39	M&O	GCC M&O. GCC M&O
42:51	GCC	GCC.
42:52	M&O	Roger can you confirm low speed tracking data leaving site?
42:55	GCC	Confirmed.
42:56	M&O.	Roger. Thank you

On-station communication

43:25	M&O	H-1 minute, all recorders to flight speed.

43:31	Record	Recorders at flight speed.
43:35	M&O	9M M&O.
43:37	9M	9 metre.
43:38	M&O	Your strip chart running?
43:40	9M	That's affirmative.
43:41	M&O	Roger.

Local Operating Procedure

Table 1-2   STDN H-45 Interface (cont)

Time	Position	Sequence	Action
H-5	S-band, RF TLM Com Tech	48	Verify that all local signal sources are OFF and no RFI is present.
	OPSR/DFE	49	Confirm that all prepass checklists and calibrations are complete and all systems are configured in accordance with the SCM.
	OPSR/DFE	50	Select DLSM parameter list and confirm constants CAM (G,444). Select parameter list CAM (G,111) and notify DFE if MCC lines not ON.
H-2	OPSR/Houston Track	51	Houston Track verify LSR tracking data validity.
	RF TLM/TLM	52	All decom inhibits OFF.
	OPSR/S-band	53	Carrier up at mission power unless otherwise directed by SCM.
	Com Tech	54	Enable voice in accordance with the SCM.
H-1	OPSR	55	Announce on internal loop, "One minute to AOS, all recorders to flight speed." (TV recorder included if applicable.)

Appendix 7 – Full text of all voice communications recorded by the author during the first orbital flight of the Space Shuttle.

These recordings were made during the first flight of the Space Shuttle.

The crew of the Columbia were John Young and Bob Crippen.

Timeline of passes for the first Space Shuttle flight tracked at Orroral.

(Z time is GMT or UTC,

MET is Mission Elapsed Time taken from the time of launch)

Launch 12 April 1981 12:00Z,   13 April 1981 01:00 local time, 00:00 MET

Orroral Tracking Passes (approximate times)

Orbit 1 13:01Z 13 April,   00:01 local time 14 April,   01:01 MET

Orbit 2 14:35Z 13 April,   01:35 local time 14 April,   02:35 MET

Orbit 3 16:09Z 13 April,   03:09 local time 14 April,   04:09 MET

Orbit 14 08:14Z 14 April,   19:14 local time 14 April,   20:14 MET

Orbit 15 09:55Z 14 April,   20:55 local time 14 April,   21:55 MET

Orbit 16 11:21Z 14 April,   22:21 local time 14 April,   23:21 MET (Waltzing Matilda)

Orbit 17 12:53Z 14 April,   23:53 local time 14 April,   24:53 MET

Orbit 18 14:28Z 14 April,   01:28 local time 15 April,   26:28 MET

Orbit 30 08:10 15 April   19:10 local time 15 April,   44:10 MET

Orbit 31 09:48 15 April   20:48 local time 15 April,   45:48 MET

Orbit 32 11:21 15 April   22:21 local time 15 April,   47:21 MET

Orbit 33 12:57 15 April   23:57 local time 15 April,   48:57 MET

Orbit 34 14:18 15 April   01:18 local time 16 April,   50:18 MET

Passes recorded on tape:

       Orbit 2

       Orbit 14

       Orbit 15

       Orbit 16

       Orbit 17

       Orbit 18.

The tape is a recording of three communications circuits recorded on a single-track recorder. Therefore, there are a few occasions where conversations overlapped and could not be properly deciphered. The communications circuits recorded were:

1. The off-station communications between Orroral, the control centres in the United States, and other stations (External circuit.)

2. The air to ground number one circuit between Orroral and the MCC Houston (Air/ground circuit.)

3. The internal on-station communications circuit within the Orroral Valley Tracking Station.

As far as possible this transcript follows the actual conversations on all of these circuits in correct time sequence.

Most callsigns are self-explanatory, but where necessary definitions have been given.

TAPE PART 1

Tape transcript of orbit 02 at approx. 01:35 local time 14 April 1981.

(The Yarragadee station was a voice only communications station in Western Australia, located a little south of Geraldton, and supervised by Orroral. CAP COM is the Houston astronaut voice communicator, the only person authorised to speak to the space shuttle except in an emergency.)

Tape Time (Min/sec)	Callsign	Transcript.
00:29	CAP COM	Hello Columbia, talking to you through Orroral, we have you for 4 minutes.
	SPACE SHUTTLE	Okay Dan. We can see some big cities down there.
	CAP COM	Well good.
	SPACE SHUTTLE	Yeah, See all the lights. We're just now locking up on our stuff here.
01:07	CAP COM	Roger. Man we locked onto those like we knew what we were doing.
	CAP COM	That's good.
01:25	CAP COM	And Columbia PLT we'd like to go do this DFI PCM procedure if you're available.
	SPACE SHUTTLE	We are available.
	CAP COM	Roger I'd like to go back on Panel R-11 and take the DFI PCM forward container circuit breaker to close, that's the one you opened earlier.
	SPACE SHUTTLE	OK, it's closed.
	CAP COM	And then the DFI PCM recorder to stop and then to forward control.
	SPACE SHUTTLE	Alrighty
	CAP COM	OK and we'd like you to follow ....
	SPACE SHUTTLE	We've got an awful lot of static in the comm here and I....
	CAP COM	Roger, you if got that to stop and to forward control, we'd like to go to Panel C-3 and take the DFI PCM recorder.......
02:16	SPACE SHUTTLE	If you can read me, you are completely unreadable at this time.
	CAP COM	Roger, Columbia.
02:35	CAP COM	Columbia, Houston, how do you read now?
	SPACE SHUTTLE	Read you ok, Dan.

	CAP COM	Roger, we'd like to press on with this procedure if you're reading me ok. I wanted the DFI PCM recorder to stop and then to forward control and following that to go to Panel C-3 and take the DFI PCM recorder to high sample.
03:32	SPACE SHUTTLE	Ok, Dan, I was getting a little bit of static and still am. You want me to be in stop on the recorder which I am at, you want me to go forward control and then go to high sample, is that affirmative?
	CAP COM	Columbia, that is affirmative.
	SPACE SHUTTLE	Ok, we go to forward control at this time and we are in high sample at this time and our talk back has remained great.
	CAP COM	Roger, we're looking at it Columbia.
	SPACE SHUTTLE	I don't know if you can see these torqueing angles Dan, but they look great. The biggest one is on IMU3 (garbled) +05 and minus 06.
	CAP COM	Roger, we see it, John, it looks great and ...Columbia, Houston, we'd like you to pull that circuit breaker — that main C circuit breaker on the DFI PCM Forward container again, we're 20 seconds to LOS. We might see you at Hawaii at 2 plus 55, if not it will be at the States at 3 plus 01
	SPACE SHUTTLE	OK, on the recorder, it's pulled and I didn't get the AOS time we'll see wherever it is.
	CAP COM	Roger, possibly Hawaii at 2 plus 55 and the States at 3 + 01.
05:02	SPACE SHUTTLE	Ok.

(GO is Goddard Operations Controller, MILA is Merritt Island tracking station in Florida USA. TULA is the Tula Peak tracking station in New Mexico, USA.)

Tape Time (Min/sec)	Callsign	Transcript/ Communications on external circuit.
05:13	YAR	Yarragadee has LOS for voice
05:28	GO	MILA ops Goddard Ops, site co-ord.
06:30	MILA	MILA
05:31	GO	Roger MILA, you will acquire in mode 17. That will be turned on near LOS at TULA and sometime during your pass you'll run out of mode 17 you'll go to mode 13.
05:47	MILA	Roger, there will be no mode 7?
05:51	GO	Stand by one, let me check that again.
05:54	MILA	Roger.
		(Track is Houston tracking controller.)
05:55	Track	Orroral track.

05:57	Opsr	Orroral
05:50	Track	Roger, I see your 3-way data now thank you.
06:22	GO	MILA Ops Planner site coord.
06:23	MILA	MILA.
06:25	GO	Yeah, that is the entire SSME dump sequence we've got on the SCM, and we don't know when they're going to start and stop that or how much of it you will get, but when they finish that SSME dump sequence then you will go to a mode 13 dump.
06:40	MILA	Roger, we copy.

(REC: is Orroral receiver technician, USB is Orroral unified S-Band technician, DH is Orroral data handling technician, Opsr is Orroral Operations Supervisor, Peter Uzzell. Except where noted as GCC Orroral COMTECH is Philip Clark. RTC is Houston Real Time Command controller, DFE is Houston Data Flow Engineer. NST is Goddard Network Support Team.

(The callsign 'M&O' was used by the on-station operations controller to commemorate the callsign of the station operations controllers that were originally used only at the earlier manned spaceflight tracking stations.)

Tape Time (Min/sec)	Callsign	Transcript/ Communications on external circuit.
06:46	REC	Orroral has acquisition OD.
07:04	REC	Orroral has acquisition FM
07:18	REC	Orroral acquisition DFI
07:28	USB	Orroral is go for command.
07:30	RTC	RTC copies.
07:34	USB	Orroral has initiated ranging.
07:44	REC	FM dump on.
07:49	DH	PCM has lock on FM dump, 960 reverse.
08:02	Opsr	Orroral range acq is complete, delta range is 7 microseconds.
08:44	REC	FM dump off, and on again.
09:08	Opsr	DFE Orroral.
09:09	DFE	Go ahead.
09:10	Opsr	Be advised that there was a short break in the dump of approx. 2 seconds, it's back on.
09:22	HOUSTON COMTECH	Orroral Valley comtech, Houston, site coord.
09:25	ORRORAL COMTECH	Orroral comtech.

09:27	HOUSTON COMTECH	Roger, crew is complaining that they can't read us on the uplink.
09:31	ORRORAL COMTECH	Roger, uplink sounds good.
09:35	HOUSTON COMTECH	Roger.
09:43	ORRORAL COMTECH	Houston comtech Orroral comtech I confirm that uplink signal quality is good.
09:48	HOUSTON COMTECH	Roger thank you.
09:48	Opsr.	DFE Orroral go ahead.
09:53	DFE	Yes, Orroral could you have your operator give me the time the FM dump mod came on?
09:56	REC	Roger, 14:35:14.
10:00	DFE	Thankyou.
11:07	REC	FM dump off 38:27.
11:10	DFE	Thank you.
11:45	DFE	Orroral DFE.
11:46	Opsr	Orroral.
11:47	DFE	We just had a drop on the OD and it came right back in, do you have any indications what caused that?
11:53	Opsr	Standby
11:57	Opsr	We did not see anything on our decom, we held lock all the way through.
12:02	DFE	OK. Copy that, thank you.
12:09	Opsr	we are approaching LOS at this time.
12:11	DFE	DFE copies.
12:12	REC	LOS DFI, 39:30.
12:19	DFE	DFE copies.
12:38	DFE	Orroral DFE is that final LOS?
12:38	REC	LOS OD.
12:45	Opsr	DFE Orroral that is final LOS.
12:47	DFE	Thankyou.

END OF ORRORAL PASS RECORDING.

There is no tape recording of Orbit 03 at approx. 03:09 local time 14 April 1981

Tape transcript of orbit 14 with full H-45 interface. Pass at approx. 19:14 local time 14 April 1981

Tape Time (Min/sec)	Callsign	Transcript/ Communications on air/ground circuits.
13:06	NST	Orroral comtech NST comtech air to ground airground 1.
13:09	ORRORAL COMTECH	Orroral comtech airground 1.
13:11	NST	I read you 5 by, how me?
13:14	ORRORAL COMTECH	Roger you're 5 by.
13:15	NST	OK ready for keying & modulation?
13:19	ORRORAL COMTECH	That's affirmative.
13:20	NST	Alright. Garble. (Tones only, six keys)
13:37	ORRORAL COMTECH	NST air/ground keying was 100%.
13:42	NST	Roger, modulation check. (With tones) NST testing on air to ground 1, One, two, three four, five, four, three, two, one, test off.
14:05	ORRORAL COMTECH	NST air/ground Orroral comtech, keying was 100%, voice & modulation were go.

Communications on external circuit.

14:05	NST	Orroral what is your status for your upcoming H-45?
14:16	Opsr	Our status is green,
14:16	NST	Roger, I copy, stand by for air ground 2
14:18	ORRORAL COMTECH	Roger.
14:18	NST	Roger sir I would like to confirm you have completed sequence 1 at this time.
14:21	Opsr	That is affirmed.
14:21	NST	Voice, NST air to ground on air to ground checkout.
14:22	NST	I would like to confirm you are in receipt of acq data and SCM.
14:26	Opsr	Affirm.
14:28	NST	and confirm that your computer systems are loaded with SSIs 169 telemetry, 107 PCM, 82 command, and 133 track.
14:34	NST	Voice, NST air to ground on air to ground checkout
14:41	Opsr	Copy, that's affirmed on all.
14:44	NST	Ok sir, why don't you proceed with sequence 3 and advise me when it is complete.
14:49	Opsr	Sequence 3 is complete.
14:52	NST	And confirm destination code 064.

14:55	Opsr	We have a 064.
14:53	VC	Voice
14:55	NST	Roger voice can I get Orroral's air to ground 2 at this time.
14:56	VC	Roger.
14:57	NST	OK standby for NST data.
14:58	VC	You have Orroral's air to ground 2.
14:59	ND	Orroral NST data.
15:01	Opsr	Orroral
15:02	NST	Orroral comtech.
15:02	NST	Yes sir are you ready to (garble)
15:05	Opsr	That's affirmed.
15:07	NST	OK Orroral go ahead and enable the MCC lines and SSMs please

On-station communication

15:21	M&O	Computers M&O
15:23	COMP	Computers
15:24	M&O	Roger have you started logging yet?
15:26	COMP	That's negative, I'll do that now.
15:27	M&O	Righto, Thankyou

Communications on external circuit

15:38	Opsr	NST data Orroral.
15:40	NST	NST data.
15:41	Opsr	Sequence 8 is complete.
15:43	NST	Roger Orroral, we see your SSMs. Standby for NST SCE & NST telemetry.
15:49	NST	Orroral NST SCE.
15:50	Opsr	Orroral.
15:53	NST	OK would you re-initialise your SCE at this time.
15:56	Opsr	Wilco.
16:03	NST	Orroral NST telemetry. Proceed with sequences 12 & 13 and advise telemetry when ready.
16:08	Opsr	Roger.

On-station communication

16:10	M&O	Receivers M&O
16:11	REC	Receivers
16:12	M&O	Roger do you have all links set up?
16:14	REC	All links are set up minus 90 dbm.
16:18	M&O	Roger copy. Data handling M&O.
16:19	DH	Data handling.
16:19	M&O	are you locked on all 3 decoms?
16:21	DH	That's affirmative.
16:25	M&O	Roger.
16:28	M&O	Data handling M&O remove inhibits from the DFI decom only please.
16:33	DH	Inhibits removed.

Communications on external circuit

| 16:41 | NST | Roger Orroral, we see your SCE re-initialised would you configure for low data rate uplink. |
| 16:46 | Opsr | Copy low data rate. |

On-station communication

| 16:51 | M&O | SCE M&O configure for low data rate uplink. |
| 16:51 | SCE | Roger. |

Communications on external circuit

| 16:49 | NST | Orroral NST telemetry we see good DFI. That completes sequence 14, proceed with sequence 15 & 16 copy when you are ready. |
| 16:59 | Opsr | Roger |

On-station communication

| 17:00 | M&O | Data handling M&O, inhibit your DFI decom please. |
| 17:04 | DH | Inhibit is on. |

Communications on external circuit

| 17:07 | NST | Roger Orroral we see low data rate uplink, Err, We do not see DMS lock. |

On-station communication

| 17:19 | Comtech | M&O comtech |

17:21	M&O	M&O
17:21	Comtech	Yes. Can we have the uplink back please you took it away in the middle of the air/ground checks.
17:27	M&O	Stand by.
17:31	Comtech	High data rate.
17:32	M&O	NST has requested low data rate. We will be going back to high data rate shortly.
17:39	Comtech	Er, OK, well NST has asked me for high data rate on the air/ground checks. Apparently, they are not synchronised at their end. However, it did go in the middle of our check and made a mess of it.
17:54	M&O	OK
17:59	M&O	Data handling M&O.
17:59	DH	Data handling
18:00	M&O	Can you re- configure the DMS for low data rate.
18:06	DH	It was configured for high data rate. It's now configured for low data rate. Do you want it back to high data rate or stay in low?
18:12	M&O	Is it locked up?
18:14	DH	It's locked er, I've got bit sync lock but no frame sync lock.

Communications on external circuit

18:20	NST	Orroral NST SCE
18:23	Opsr	Orroral.
18:23	NST	Ok you have a problem with DMS lock?
18:27	Opsr	OK we're not showing lock at the time, there was a conflict of interest between yourselves and the air/ground people. Stand by please.
18:39	Opsr	Er, We'd like to re-initialise the SCE MUX if we may please.
18:46	NST	Roger.

On-station communication

18:49	M&O	SCE M&O
18:50	SCE	SCE
18:51	M&O	Roger, re-initialise your SCE MUX.
19:17	SCE	SCE MUX is re-initialised.

Communications on external circuit

19:17	NST	OK Orroral. We see lock on low data rate. Will you perform SCE loop test?

19:37	Opsr	NST SCE Orroral.
19:39	NST	NST SCE
19:40	Opsr	Loop is good, PEP is good, all the flags are zeros.
19:44	NST	Roger, reconfigure for high data rate uplink.
19:48	Opsr	Wilco.

On-station communication

19:49	M&O	SCE M&O, reconfigure for high data rate uplink.
19:51	SCE	Wilco.
19:52	M&O	Data handling M&O.
19:54	DH	Data handling.
19:55	M&O	Reconfigure for high data rate.
19:57	DH	Reconfigured.
20:04	DH	We have lock.
20:06	SCE	SCE is reconfigured.
20:06	M&O	Roger.
20:07	M&O	Data handling M&O, remove your inhibits on the OD & FM dumps please.
20:15	DH	Inhibits removed.
20:16	M&O	Roger.

Communications on external circuit

20:22	Opsr	NST SCE Orroral,
20:24	NST	NST SCE.
20:25	Opsr	We are configured in high data rate.
20:28	NST	OK you can perform SCE loop test.
20:35	Opsr	NST telemetry Orroral.
20:38	NST	Telemetry.
20:39	Opsr	16 is complete.
20:42	NST	Roger, stand by one and we will verify.
20:43	Opsr	NST SCE Orroral.
20:45	NST	SCE
20:46	Opsr	Loop test on high data rate, loop is good, PEP is good, flags are zeros.
20:51	NST	Roger, safe the SCVM.

20:55	Opsr	SCVM is safe
20:58	NST	Roger, we see it.
21:10	NST	Orroral your data looks good for telemetry. That completes sequences 18 er 17 & 18. Proceed with sequence 19 and stand by for data 20.
21:21	Opsr	Wilco

On-station communication

21:21	Comtech	M&O comtech
21:23	M&O	M&O.
21:24	Comtech	Roger. We have completed air/ground tests, we have a go on the air/ground interface.
21:29	M&O	Roger, copy

Communications on external circuit

21:29	NST	Orroral NST data.
21:30	Opsr	Orroral
21:32	NST	Roger. Disable your MCC lines. Disable NOCC lines enter & verify destination code 160.
21:39	Opsr	Roger

On-station communication

21:30	M&O	Data handling M&O.
21:32	DH	Data handling.
21:33	M&O	Inhibit all decoms.
21:36	DH	Inhibited.
21:48	M&O	Data handling M&O
21:49	DH	Data handling.
21:50	M&O	Configure SOU2 for format 151 please.
21:54	DH	Stand by.
21:58	USB	M&O USB
22:00	M&O	M&O
22:01	USB	Can I take carrier down?
22:04	M&O	That's affirmative.
22:05	USB	Carrier is down, modulation is off.

22:07	M&O	Roger.
22:09	DH	SOU configured.
22:10	M&O	Thank you
22:11	USB	M&O USB.
22:12	M&O	M&O.
22:14	USB	Er, give the boys up here a good turn around.
22:19	M&O	(garble) Timmy. Er, sometime later we'll bear it in mind.
22:25	USB	Roger thank you.
22:35	M&O	Data handling M&O.
22:37	DH	Data handling.
22:37	M&O	Roger. Is your MSFTP-3 locked up now?
22:40	DH	That's affirmative.

Communications on external circuit

22:41	Opsr	NST data Orroral.
22:43	NST	NST data.
22:45	Opsr	MCC lines are off. Destination code is a 160.
22:40	NST	Roger Orroral, stand by for data Com.
22:53	Goddard Ops	Orroral Goddard Ops.
22:54	Opsr	Orroral.
22:55	Goddard Ops	Alright sir we'll give you a go on you NOCC interface. Stand by and we'll put you on the site coord for your MCC interface.
23:01	Opsr	Roger
23:07	Goddard Ops	Voice control, Goddard Ops NOCC interface.
23:11	VC	Voice.
23:11	Goddard Ops	Put Orroral to site coord please.
23:13	VC	Roger.

On-station communication

24:17	USB	M&O USB
24:19	M&O	M&O.
24:20	USB	On sequence 35 we show dummy load. Do you require dummy load or do you want antenna for that one?
24:26	M&O	Stand by.

Communications on air/ground circuits

24:30	HOUSTON COMTECH	Orroral Valley comtech Houston air to ground one.
24:33	ORRORAL COMTECH	Orroral comtech air/ground 1.
24:35	HOUSTON COMTECH	You're 5 by. Meet me on air to ground 2.
24:28	ORRORAL COMTECH	Roger.

Communications on external circuit

24:33	RTC	Orroral RTC.
24:34	Opsr	Orroral.
24:36	RTC	Configure for a H-30 interface
24:44	Opsr	Wilco.

On-station communication

24:50	M&O	USB M&O
24:51	USB	USB.
24:52	M&O	Roger are you configured for sequence 35?
24:55	USB	That's affirmed, all mode, carrier up.
24:59	M&O	Roger.
25:00	USB	Roger, we are configured.

Communications on external circuit

| 25:01 | Opsr | RTC Orroral. |
| 25:04 | RTC | RTC. |

On-station communication

| 25:03 | USB | We are to operate. |
| 25:04 | M&O | Roger that. |

Communications on external circuit

| 25:06 | Opsr | We are configured per sequence 35. |
| 25:10 | RTC | Roger. |

Communications on air/ground circuits

| 25:14 | HOUSTON COMTECH | Orroral Valley Houston you've got the teleprinter coming at you now. |
| 25:19 | ORRORAL COMTECH | Roger. |

On-station communication

25:31	Comtech	GCC comtech on link 2
25:33	GCC	GCC

Communications on external circuit

25:34	RTC	Orroral RTC.
25:35	Opsr	Orroral.

On-station communication

25:34	Comtech	Roger. Could you confirm the frequency and level of that signal?

Communications on external circuit

25:37	RTC	We have a good interface, you may safe your SCVM.
25:41	Opsr	Copy.

On-station communication

25:47	USB	SCVM safe.
25:51	USB	M&O USB can I (garble).
25:43	M&O	Stand by one.

Communications on air/ground circuits

25:51	GCC	Break, break,
25:57	GCC	Houston comtech Orroral comtech. We have a frequency of 2056 a level of negative 12.
26:06	HOUSTON COMTECH	Negative 12 at the comtech console?
26:10	HOUSTON COMTECH	What have you got going into your DMS?
26:12	ORRORAL COMTECH	Roger, standby we are just setting that now.
26:15	HOUSTON COMTECH	OK

On-station communication

26:04	M&O	USB M&O loop (garble)
26:08	USB	Roger
26:44	Comtech	GCC comtech link 2.
26:49	GCC	GCC.
26:49	Comtech	Roger are you letting that tone come through to the console. I can't hear it.
26:52	GCC	No we broke it.

26:55	Comtech	OK, I had better ask him to put it back I didn't hear him break it. I hadn't finished setting the level here.
27:01	GCC	Well I told him to break when I read the frequency and the level of it.

Communications on air/ground circuits

27:06	ORRORAL COMTECH	Houston comtech Orroral comtech.
27:09	HOUSTON COMTECH	Go ahead.
27:09	ORRORAL COMTECH	Roger, could you put the tone back? We had not completed setting the level when it was taken off.
27:15	HOUSTON COMTECH	OK, you got it again.
27:17	ORRORAL COMTECH	Roger thank you. Stand by please.
27:19	HOUSTON COMTECH	And I do need the verification level also.
27:23	ORRORAL COMTECH	Roger that. Stand by.
28:04	ORRORAL COMTECH	Houston comtech Orroral comtech
28:06	HOUSTON COMTECH	Go ahead.
28:07	ORRORAL COMTECH	The level at the DMS is neg 22, the level at the verification is neg 20.

Communications on external circuit

28:14	Opsr	That is affirmed.
28:16	DFE	Madrid DFE OK Let me turn on your lines.

Communications on air/ground circuits

28:20	HOUSTON COMTECH	Ok we'll go ahead and check the S-Band voice on air to ground 1 and you can stay configured with the teleprinter on air/ground 2
28:30	ORRORAL COMTECH	Roger.
28:33	HOUSTON COMTECH	OK let me send you a mark on air to ground 1.
28:37	ORRORAL COMTECH	Roger.

Communications on external circuit

28:38	DFE	Madrid your lines are on.
28:42	Opsr	Orroral sees lines on.
28:44	DFE	Copy that stand by just a moment.

Communications on air/ground circuits

28:50	GCC	Break, break,
28:53	GCC	Houston comtech Orroral comtech we have a frequency of 2524 level negative 8.

| 29:01 | HOUSTON COMTECH | Here is a space. |

Communications on external circuit

28:59	DFE	Orroral I would like you to proceed with sequence 25 and give me a destination code of 160 please.
29:04	Opsr	You have a 160.
29:07	DFE	And I am ready for your data generator please.
29:14	Opsr	3 data streams OK?
29:17	DFE	That would be just great.

On-station communication

| 29:09 | M&O | Data handling M&O remove the inhibits on three decoms please. |
| 29:13 | DH | Copy |

Communications on air/ground circuits

29:14	GCC	Break, break,
29:17	GCC	Houston comtech Orroral comtech we have a frequency of 2474 level negative 7.5.
29:27	HOUSTON COMTECH	OK let me send you a series of keys.

Communications on external circuit

| 29:42 | DFE | Orroral DFE. Your OD is green. We'll give you a go for support. |
| 29:59 | | I'll leave you lines on and have RTC come to you |

Communications on air/ground circuits

29:42	HOUSTON COMTECH	Orroral valley Houston.
29:44	ORRORAL COMTECH	Orroral
29:46	ORRORAL COMTECH	Keying was 100% on air/ground 1
22:49	HOUSTON COMTECH	OK we might as well check out his air to ground 2 line also if you are ready I'll send you a mark on that.

Communications on external circuit

| 30:01 | OPSR | We have done the initial interface with RTC. |
| 30:06 | DFE | OK copy that then lines are going off and I would like you to print out your DLSM. |

Communications on air/ground circuits

| 30:10 | GCC | Break, break, |
| 30:13 | GCC | Houston comtech Orroral comtech we have a frequency of 2524 level negative 11. |

30:22	HOUSTON COMTECH	OK here's a space.
30:32	GCC	Break, break,
30:35	GCC	Houston comtech Orroral comtech we have a frequency of 2474 level negative 12.
30:43	HOUSTON COMTECH	OK let me send you a series of keys here on air to ground 2
30:50	ORRORAL COMTECH	Roger.
31:02	ORRORAL COMTECH	Houston comtech Orroral comtech keying 100% on air/ground 2.
31:07	HOUSTON COMTECH	OK now I guess I had better give you keying and modulation check if you are ready.
31:12	ORRORAL COMTECH	Roger, are you going to multiaccess both?
31:15	HOUSTON COMTECH	Yes, I am.
31:24	ORRORAL COMTECH	Houston comtech Orroral comtech, go ahead.

On-station communication

31:25	M&O	Comtech M&O
31:29	Comtech	Comtech.
31:30	M&O	Do you still need carrier up?
31:32	Comtech	That's affirmative, we're checking.

Communications on air/ground circuits

31:30	HOUSTON COMTECH	(with tones) Houston comtech testing one, two, three, four, five, four, three, two, one, test out.
31:46	ORRORAL COMTECH	Houston comtech Orroral comtech keying was 100% on air/ground 1 & air/ground 2, voice is go on both air/grounds.
31:54	HOUSTON COMTECH	OK now if you would give me downlink tones
31:56	ORRORAL COMTECH	Just stand by please.

On-station communication

31:59	Comtech	M&O comtech we have finished with the uplink and data handling could you give me sense switch 4 on SOU 2, er SOU 1.
32:06	M&O	USB M&O.
32:07	USB	USB
32:07	M&O	Terminate the carrier.
32:10	USB	Roger, carrier is down, modulation os off.

Communications on air/ground circuits

32:16	HOUSTON COMTECH	Neg 12.5 on both circuits.

32:21	ORRORAL COMTECH	Roger.
32:22	HOUSTON COMTECH	OK looks good and we'll give you a go and configure for teleprinter on your air to ground 2, that is configuration Lima.
32:30	ORRORAL COMTECH	Roger, configuration Lima.
32:33	HOUSTON COMTECH	Roger that, OK thank you much.
32:35	ORRORAL COMTECH	Roger.

On-station communication

32:07	M&O	USB M&O
32:08	USB	
32:09	M&O	Terminate the carrier please
32:10	USB	Roger carrier is down modulation is off
32:16	USB M&O	(garble)
32:41	Comtech	M&O comtech.
32:46	M&O	M&O.
32:48	Comtech	Roger we have a go on our Houston interface, we have changed the configuration, we'll be operating air/ground configuration Lima for teletype.
33:00	Opsr	Also in the sleep configuration, is that right?
33:03	Comtech	I believe so.
33:04	Opsr	OK
33:06	Opsr	What, so you inhibit the QUINDAR and....
33:06	Comtech	I inhibit the QUINDAR on 1 and go with teletype on the other.
33:12	Opsr	OK
33:13	Comtech	It is a little unusual I guess, I think I had better go back and query him and make sure.
33:17	Opsr	That's affirmed.

Communications on air/ground circuits

33:25	ORRORAL COMTECH	Houston comtech Orroral comtech site coord.
33:29	HOUSTON COMTECH	Houston comtech
33:30	ORRORAL COMTECH	Roger with this configuration Lima that we have just been advised, in view of the sleep configuration on the SCM I take it that we will inhibit the QUINDAR on air/ground 1 and go with the normal teletype configuration on air/ground 2. Is this correct?
33:45	HOUSTON COMTECH	That's correct.

| 33:46 | ORRORAL COMTECH | Roger, understood. |
| 33:47 | HOUSTON COMTECH | Thank you Orroral. |

On-station communication

34:00	M&O	Recorders M&O
34:01	RECORD	Recorders.
34:02	M&O	Roger could you normalise your DSS532 recorder please, and in a couple of minutes we'll be ready for cal.
34:10	REC	OK, give me a mark but we're ready anyway.
34:12	M&O	Roger.
34:20	RECORD	And 532 is back to normal configuration.
34:23	M&O	Roger, thank you.
34:20	Comtech	M&O comtech, airground 1 keying is inhibited at this time.
34:34	M&O	Copy thank you.

Communications on external circuit

34:57	Opsr	DFE Orroral
34:59	DFE	Go ahead Orroral.
35:00	Opsr	We'd like to put the signals around the loop to do a pre-pass calibration at this time. Is that OK?
35:08	DFE	Stand by just a moment.
35:12	DFE	Say that again what you want to do Orroral.
35:14	Opsr	Just put signals around our RF receive loop for pre-pass calibration.
35:19	DFE	That would be fine. Just be sure that I don't get any data generator telemetry on line.
35:24	Opsr	Wilco.

On-station communication

35:24	M&O	Data handling M&O.
35:24	DH	Data handling
35:25	M&O	Confirm all decoms are inhibited.
35:28	DH	That's confirmed.

Communications on external circuit

| 35:28 | DFE | In fact, Orroral, just to sure that I don't get it I'm going to go ahead and turn your lines off, I'll re-enable them at H-5. |

| 35:35 | Opsr | OK. |

On-station communication

35:31	9M	M&O 9 metre.
35:37	9M	M&O 9 metre.
35:38	M&O	M&O
35:39	9M	Which col tower do you want us to put this onto?
35:32	M&O	9 metre col tower.
35:34	9M	Roger.
35:46	M&O	And go to the col tower, disable both axis' and select autotrack.
35:53	9M	Roger, will do.
36:04	M&O	Receivers M&O.
36:05	REC	Receivers.
36:06	M&O	Are you ready for cals?
36:08	REC	Yes, we'll be starting on the minute.
36:12	Record	Receivers recorders.
36:15	REC	Receivers.
36:17	Record	We have our tape rolling.
36:39	9M	9 metre M&O, er 9 metre
36:47	M&O	9 meter M&O go ahead.
36:48	9M	We are on the 9 metre coll tower with axis' disabled and in autotrack.
36:55	M&O	Roger thank you.
37:12	M&O	Receivers M&O.
37:15	REC	Receivers.
37:15	M&O	As soon as you finish cals kill the RF loop please.
37:31	REC	That ends calibrations recorders.
37:34	Record	Roger, copy.
37:35	Record	And recorders back to flight speed.

Communications on external circuit

37:39	Track	Orroral track.
37:40	Opsr	Orroral
37:42	Track	Are you configured for sequence 42 of the H45 count?

37:46	Opsr	Stand by, we just have to lock up the ranging.

On-station communication

37:46	M&O	USB M&O
37:50	USB	USB
37:51	M&O	Can you set up for a zero-delay turnaround?
37:54	USB	Roger, standby
38:22	USB	M&O USB we are configured.

Communications on external circuit

38:27	Opsr	Track Orroral.
38:28	Track	Track go ahead.
38:29	Opsr	We are configured per sequence 42.
38:32	Track	Roger you can put your data on line whenever you are ready.

On-station communication

38:36	M&O	9 metre M&O put data on line.

Communications on external circuit

38:40	Opsr	Data is on line.
38:42	Track	Copy

On-station communication

38:41	M&O	GCC M&O.
38:44	GCC	GCC
38:45	M&O	Confirm data leaving please, tracking data.
38:51	GCC	Roger, it's going now.
40:32	M&O	SCE M&O.
40:34	SCE	SCE.
40:35	M&O	Roger, we should come up shortly on a check of the H-10 sequence.

Communications on external circuit

40:38	Track	Orroral Houston track.
40:40	Opsr	Orroral.
40:42	Track	Your data looks good you can take it off line.

40:46	Opsr	Wilco

On-station communication

40:49	9M	Low speed data terminated.
40:55	USB	M&O USB can I terminate now?
41:01	M&O	Yes, you can probably take carrier down, no leave carrier up, just terminate ranging.
41:06	USB	Roger.
41:10	9m	M&O 9 metre servo.
41:12	M&O	M&O
41:12	9M	Do you wish us to go to zenith at this time?
41:16	M&O	That's affirmative.
41:18	9M	Roger.

Communications on external circuit

41:23	RTC	Configure for sequence 45 please,
41:26	Opsr	Wilco.

On-station communication

41:27	USB	SCVM to operate.
41:29	M&O	Carrier up, modulation on.
41:31	USB	That's affirmed, we left it up.
41:33	M&O	Roger

Communications on external circuit

41:34	Opsr	RTC Orroral.
41:37	RTC	RTC.
41:37	Opsr	We're configured.
42:04	RTC	Orroral RTC.
42:05	Opsr	Orroral.
42:07	RTC	We have good interface test, you may drop modulation.
42:13	Opsr	Wilco.

On-station communication

42:13	M&O	USB Modulation off please
42:15	USB	Roger, carrier is down, modulation is off.

42:23	M&O	9-meter M&O.
42:24	9M	9 meter.
42:25	M&O	You can go to intercept angles now.
42:27	9M	Intercept angles, roger.
42:31	M&O	Data handling remove decom inhibits.
42:32	DH	Inhibits off.
42:35	USB	USB carrier is up.
42:37	M&O	Roger that,
42:37	9M	Low speed data on line.
42:39	M&O	GCC M&O. GCC M&O
42:51	GCC	GCC.
42:52	M&O	Roger can you confirm low speed tracking data leaving site?
42:55	GCC	Confirmed.
42:56	M&O	Roger. Thank you.

Communications on external circuit

43:01	DFE	Quito DFE.
43:06	Q	This is Quito.
43:07	DFE	I understand we have you for an H minus 30, is that correct?
43:11	Q	That's correct.
43:12	DFE	OK, I'm coming up on a pass with Orroral, and we'll be done at about 08:17 so about 08:20 I should be ready to do the H minus 30 interface with you.
43:22	Q	Copy.

On-station communication

43:25	M&O	H-1 minute, all recorders to flight speed.
43:31	Record	Recorders at flight speed.
43:35	M&O	9M M&O.
43:37	9M	9 metre.
43:38	M&O	Your strip chart running?
43:40	9M	That's affirmative.
43:41	M&O	Roger.

Communications on external circuit

44:32	REC	Orroral has acquisition OD.
44:36	DFE	DFE copies.

On-station communication

44:38	DH	M&O data handling we have lock on OD & DMS.
44:41	M&O	copy thank you

Communications on external circuit

44:42	USB	Orroral is go for command.
44:42	REC	Orroral has acquisition of DFI.
44:44	DH	PCM lock on DFI.
44:46	RTC	RTC copies.
44:53	USB	Ranging is initiated at Orroral.

On-station communication

44:56	9M	We have valid autotrack.

Communications on external circuit

44:59	REC	Acquisition FM.

On-station communication

45:02	M&O	Any modulation on the FM?

Communications on external circuit

45:06	REC	No modulation on FM.
45:09	REC	Modulation on FM
45:12	DFE	DFE copies.
45:13	DH	PCM has lock on FM dump, 1024 reverse

On-station communication

45:16	USB	M&O USB
45:22	9M	M&O 9 metre.

Communications on external circuit

45:22	Opsr	Range acq is complete, range delta is 6.6 microseconds.

On-station communication

45:24	M&O	9 Meter.
45:24	9M	The program is nominal.
45:27	M&O	say again 9 metre
45:32	9M	The program is nominal.
45:34	M&O	Roger thank you, no bias?
45:37	9M	No time bias.
45:38	M&O	Why do we have a low servo bandwidth?
45:41	9M	We're skimming the horizon at the moment.
45:44	M&O	OK

Communications on external circuit

45:48	DFE	Orroral DFE.
45:50	Opsr	Orroral.
45:51	DFE	We just had a drop on the OD telemetry. Do you have any indications in house?

On-station communication

45:57	9M	It should improve now, Hugh. We are above the horizon now, we have left the horizon now.

Communications on external circuit

45:59	Opsr	We are on the horizon. Stand by please
46:02	9M	We are above the horizon now, we have left the horizon now.

On-station communication

46:05	M&O	Data handling M&O did you see any drops on the OD decom?
46:07	DH	The signal got pretty weak. Yes I think we did have a drop but I was watching the DMS at the time.

Communications on external circuit

46:15	Opsr	DFE Orroral.
46:16	DFE	Orroral DFE
46:17	Opsr	OK we did see a couple of rough spots, we were walking along the top of the hill.
46:21	DFE	OK thank you.
46:24	OPSR	We are clear of the horizon at this time.

On-station communication

46:24	DH	M&O PCM
46:26	M&O	M&O
46:26	DH	Roger, we saw some drops on the FM it looks solid at this time.
46:31	M&O	Roger copy, thank you
46:40	Comtech	Data handling comtech.
46:41	DH	Data handling
46:42	Comtech	roger can you check the level of the level of the incoming to the DMS of the teletype for me?

Communications on external circuit

46:46	Opsr	DFE Orroral.

On-station communication

46:50	M&O	Data handling M&O, right can you dial up that word take a count of the number of drops that we have on this FM dump?
47:02	DH	Roger.
47:02	M&O	We want to know at the end, not right now.

Communications on external circuit

47:06	Opsr	DFE Orroral.
47:12	DFE	Orroral DFE
47:13	Opsr	Be advised we did notice a couple of drops in the dump. It was approximately 1 minute ago.
47:19	DFE	OK copy, and do you know how long the duration was of the drop?
47:23	Opsr	They were only momentary.
47:26	DFE	Copy that.
48:35	RTC	Quito RTC
48:37	Q	Quito.
48:39	RTC	Configure for a command interface test per sequence 35 please.
48:42	Q	Wilco.
48:50	HOUSTON COMTECH	Orroral comtech Houston comtech
48:53	ORRORAL COMTECH	Orroral comtech.
48:55	HOUSTON COMTECH	Ok teleprinter message is completed you can configure per the SCM.

49:00	ORRORAL COMTECH	Roger. We still have the 450 Hz downlink.
49:04	HOUSTON COMTECH	Thank you.
49:06	Q	RTC Quito,
49:08	RTC	RTC
49:09	Q	Ready for sequence 35.
49:11	RTC	Roger.
49:15	ORRORAL COMTECH	Houston comtech Orroral comtech.
49:19	HOUSTON COMTECH	Go ahead.
49:21	ORRORAL COMTECH	Air/ground 2 is configured per the SCM, 450 Hz tone has gone off the downlink.
49:25	REC	Dump off at Orroral 16:45.
49:27	DFE	Thankyou.

On-station communication

49:30	DH	M&O PCM
49:31	M&O	M&O
49:32	DH	FM drop lock,
49:34	M&O	Roger.

Communications on external circuit

49:36	RTC	Quito RTC.
49:37	Q	Quito.
49:39	RTC	We have a good interface test, You can safe your MUX.

On-station communication

49:41	M&O	Data handling M&O

Communications on external circuit

49:43	Opsr	Orroral has loss of dump data.

On-station communication

49:43	DH	Data handling
49:35	M&O	How many drop locks did we have?

Communications on external circuit

49:46	DFE	DFE copies Orroral.
49:48	Opsr	RTC Orroral

49:50	REC	FM off 17:00
49:51	RTC	Go ahead Orroral
49:52	Opsr	You have about 20 seconds to horizon.
49:55	RTC	Copy.

On-station communication

50:22	DH	M&O PCM.
50:23	M&O	M&O
50:24	DH	Ah yeah, we dialled the number of errors here we found that it's too much. I suspect that we dropped lock completely then re-acquired again and I think that's why we have so many errors.

Communications on external circuit

| 50:31 | REC | DFI off 17:48. |
| 50:36 | DFE | DFE copies. |

On-station communication

| 50:40 | DH | Loss of lock on OD. |
| 50:44 | DH | The data went completely at one stage and then we re-acquired again. |

Communications on external circuit

| 50:47 | DFE | Is that final LOS on the OD Orroral? |
| 50:50 | REC | Not Yet. |

On-station communication

| 50:50 | M&O | What your saying is we have got a full register. |
| 50:54 | DH | Yeah, the display shows 146. |

Communications on external circuit

50:58	REC	LOS on the OD 18:15.
51:02	Opsr	That's LOS all links.
51:04	DFE	DFE copies.
51:18	Opsr	DFE Orroral.
51:20	DFE	Orroral DFE.
51:21	Opsr	Telemetry advises that there were 146 dropouts in the dump.
51:28	DFE	146 dropouts, thank you very much.

On-station communication

| 51:59 | Comtech | Air/ground circuits released. |
| 52:01 | USB | Low speed data terminated. |

Communications on external circuit

| 52:01 | DFE | Orroral DFE, your lines are off would you configure for post-pass. |
| 52:05 | Opsr | Wilco. |

On-station communication

52:06	USB	S-Band carrier down.
52:08	DH	Inhibits on.
52:12	M&O	Recorders M&O
52:13	Record	Recorders.
52:14	M&O	Stop recording and re-configure for playback please.
52:16	Record	Wilco.
		END OF ORROAL PASS RECORDING
52:38		Wake up music played to Shuttle through Quito. "Blast-off Columbia" Sung by Roy McCall.

Communications through Quito station.

54:18	CAPCOM	Hullo Columbia, welcome to day 2
54:22	SPACE SHUTTLE	All right. Good morning, gents, how is the silver team this morning?
54:29	CAPCOM	Well, we are just fine. We had a grand night. Things are looking good and we do have a question. We were wondering if you guys are shivering up there or is the temperature pretty good?
54:39	SPACE SHUTTLE	Well, it certainly got a little chilly last night. I was ready to break out the long undies. If you guys have got a way to warm up the cabin a little bit, we would probably be interested in hearing about it? Also, for the GAP, I do not know if he noticed when we came over the hill there, but apparently, I did not do my item two before I was through, we did not get to freeze until I dumped so I had it coming down again.
55:23	CAPCOM	Roger Columbia, stand by a second.
55:35	CAPCOM	Columbia Houston, we think that we took the recorder away from you and you will probably have to do it again.
55:32	SPACE SHUTTLE	Ok, well I am all set up to do that. We will just get it after we go LOS then.
55:40	CAPCOM	ROGER, that will be great.
		END OF QUITO PASS RECORDING

TAPE PART 2

Tape Transcript Orbit 15 at approx. 20:49 local time 14 April 1981

Tape Time (Min/sec)	Callsign	Transcript
		Communications on air/ground circuits
00:13	ORRORAL COMTECH	Orroral Comtech airground
00:15	HOUSTON COMTECH	Stand by one
00:26	HOUSTON COMTECH	Comtech Houston comtech airground two.
00:30	ORRORAL COMTECH	Orroral Comtech airground two,
00:33	HOUSTON COMTECH	Standby for H-5 check
00:36	ORRORAL COMTECH	Houston comtech Orroral comtech we've got a very bad echo.
00:41	HOUSTON COMTECH	Roger, copy bad echo on this circuit?
00:43	ORRORAL COMTECH	That's affirmative and we had it on airground one as well.
00:50	HOUSTON COMTECH	Standby one.
01:24	VC	Orroral voice how do you copy?
01:27	ORRORAL COMTECH	Voice Orroral you're loud & clear, no echo.
01:29	VC	Roger.
01:55	HOUSTON COMTECH	Orroral comtech Houston comtech airground two.
01:59	ORRORAL COMTECH	Houston comtech Orroral comtech airground two you're loud & clear, no echo.
02:02	HOUSTON COMTECH	Okay, standby.
02:05	ORRORAL COMTECH	Roger.
02:06	HOUSTON COMTECH	Houston comtech testing, (using tones) One, Two, Three, two one, end of test.
02:17	ORRORAL COMTECH	Houston comtech Orroral comtech keying 100% on airground one and airground two.
02:22	HOUSTON COMTECH	Roger
		Orroral & Yarragadee communications on external circuit.
02:13	Opsr	RTC Orroral.
02:16	RTC	Orroral RTC

02:17	Opsr	we are configured for uplink interface.
02:24	Y	Houston comtech Yarragadee comtech we have 100% keying & modulation go.
02:28	HOUSTON COMTECH	Roger, configure for your pass.
02:31	ORRORAL COMTECH	Orroral, roger.
02:38	Y	Yarragadee is configured for real-time support.
02:42	HOUSTON COMTECH	Roger Yarragadee.
02:49	RTC	Orroral RTC,
02:50	Opsr	Orroral
02:51	RTC	We have a good command interface you can drop modulation.
02:55	Opsr	Wilco.

Yarragadee talking on external circuit.

| 04:38 | Y | Yarragadee is go for voice. |

SPACE SHUTTLE is through Yarragadee.

04:42	CAP COM	Hullo Columbia, we're talking to you through Yarragadee, we have you for 7 ½ minutes.
04:51	?	Go ahead MILA
04:53	GO	Mila Goddard ops site coord
04:56	SPACE SHUTTLE	Load & clear there Houston.
04:58	CAP COM	Roger, & I have the pad for your RCS test sequence number one on 2 dash 42 of your cap.
05:33	SPACE SHUTTLE	OK Dan I'm ready to copy.
05:37	CAP COM	Roger, on the burn attitude Roll 179, pitch 164, yaw 320 ;the target HA 145 HP is plus 144, delta V total is 0001.8, Tgo is 3 seconds, down in the notes as Plus x trans check that box I'm going over to item 21 is 207600, item 27 TIG is 000/222000.0, item 36 - 0001.8, 37 plus all zip, 38 is all zip, and the post burn attitude is NA.
06:57	SPACE SHUTTLE	OK I'll read back are as follows burn attitude is 179164320, 145 by 144. 1.8 3 seconds 207600 is the weight 22 hours 20 minutes no seconds -1.8 all zip all zip post burn attitude not applicable and it's a plus X translation.
07:28	CAP COM	Roger and interconnect the note that the interconnect to RCS from the left OMS.
07:38	SPACE SHUTTLE	OK. We are in that configuration right now.
07:31	CAP COM	Roger.

Communications on external circuit

| 07:46 | Opsr | DFE Orroral. |

07:50	DFE	Orroral, DFE
07:52	Opsr	Do we have a clearance for destination code 150?
07:54	DFE	That's affirmative, I have turned your lines on, 150 at this time please.
08:03	Opsr	You have it.
08:08	DFE	Roger I see it and would you confirm you are configured for your upcoming support?
08:12	Opsr	We are configured.

SPACE SHUTTLE is through Yarragadee.

08:04	SPACE SHUTTLE	OK Daniel I have a couple for you.
08:08	CAP COM	Roger we are ready to copy.
08:11	SPACE SHUTTLE	OK as you can see I've got the fuel cells purge going right now. When I was doing the heater reconfig, I discovered down on ML 86 bravo that we had both water line heaters circuit breakers closed so I have opened A and we are running on bravo only. I don't know whether you will want to consider opening changing those around to verify that alpha is working later.
08:37	CAP COM	Roger, we copy.

Communications on external circuit

08:44	LRCO	GC LRCO on site coord
08:46	GC	LRCO GC.
08:49	LRCO	Yes Sir, Er, We were just talkin' here we have of course landing opportunities at KSC on orbits 16 through 21 and SR test 3733 and to Edwards on 18 through 23 SR test 6108. Er normally we'd go though a full minus count for those landing opportunities. We'd like to get (garble) unless you have some intentions to land on those opportunities from conducting our complete five hour minus count.
09:27	GC	OK I'll tell you what we have. The GC who has been working that is due here in (garble)
09:15	CAP COM	And Columbia Houston can we get an alignment report.

Communications on external circuit

| 09:48 | LRCO | Seems a little foolish to come up and go through a full minus count if you don't have any intentions to land. We would still have the resources available if it certainly became necessary to land. |
| 09:57 | GC | Roger |

SPACE SHUTTLE is through Yarragadee.

| 09:34 | SPACE SHUTTLE | OK the torqueing angles were -.08 + .21 -.17 -.23 .05 .01 -.20 - .06 -.13 the time was 21-32-52 and star error was 200 so I guess that that is a good angle. |

10:16	CAP COM	Roger, we copy.
10:21	SPACE SHUTTLE	That was IMU's 1, 2, and 3 respectively.
10:25	CAP COM	Roger.
10:26	SPACE SHUTTLE	X, Y & Z respectively.
10:46	CAP COM	Columbia, Houston. You broke up a little bit. We copied the time and then something 200, but we missed what was in between.
10:58	SPACE SHUTTLE	Did you get the torqueing angle?
11:02	CAP COM	Yes, we did.
11:11	SPACE SHUTTLE	All angles (garbled faint)
11:21	CAP COM	Columbia, we are not reading you. We will catch the rest of this at Orroral in about 3 Min

Yarragadee talking on external circuit.

11:50	Y	Yarragadee has 1 minute to LOS.

Communications on external circuit

12:41	REC	Orroral has acquisition OD.
12:44	DFE	DFE copies.
12:36	REC	Orroral has acquisition DFI.
12:53	Y	Yarragadee has LOS.
12:58	USB	Orroral is go for command.
13:01	RTC	RTC copies.
13:03	REC	Orroral has acquisition FM, carrier only.
13:08	USB	Orroral has initiated ranging.

SPACE SHUTTLE is through Orroral

13:08	CAP COM	Hello. Columbia. We are talking to you through Orroral. We have you for 5 and a half minutes.

Communications on external circuit

13:15	REC	Orroral has dump, 46:41.
13:19	DFE	DFE copies.
13:21	DH	PCM lock on dump FM, 960 reverse.
13:31	Opsr	Orroral range acquisition is complete. Delta range is 10.6 microseconds.

SPACE SHUTTLE is through Orroral

13:17	SPACE SHUTTLE	OK, Daniel.

13:19	CAP COM	Roger, our question on that—We copied the torqueing angles, we copied the execution time. However, after the execution time there' was something that came through garbled followed by 200. And that's where our question lies.
13:31	SPACE SHUTTLE	John said that they had very small angle difference and consequently we thought it had a good angle.
13:41	CAP COM	Roger, we copy that, thank you.

Communications on external circuit

14:18	DFE	Orroral, DFE.
14:21	Opsr	Orroral/
14:22	DFE	Could you give me a reading off the DFI on the tyre pressures?
14:31	Opsr	Wilco.
14:42	REC	FM modulation off, 48:05.
14:46	DFE	DFE copies.
14:49	REC	Modulation back on.
14:53	DFE	Copy that Orroral, that was swapping tracks.
14:56	REC	Roger.
16:35	Opsr	DFE Orroral.
16:37	DFE	Orroral DFE
16:39	Opsr	Tyre pressure readouts. Measurement no. 1 317 left nose. Measurement 2 334 right nose, no. 3 331 right main outboard, measurement 4 323 right main inboard, measurement 5 is 224 left main outboard, measurement 6 is 217 left main inboard.
17:21	DFE	Thank you Orroral.
17:42	Opsr	Track Orroral.
18:06	Opsr	RTC Orroral you have about 1 minute to horizon.
18:11	RTC	RTC copies.
18:22	REC	Dump off 51:37.

SPACE SHUTTLE is through Orroral

18:22	CAP COM	And Columbia Houston for your water supply dump the numbers are for alpha & bravo there will be no dump at this time.
18:31	SPACE SHUTTLE	Roger.
18:32	CAP COM	And you are 30 sec to LOS. We will see you at MILA at 22 + 23
18:41	SPACE SHUTTLE	Alright.

Communications on external circuit

18:42	REC	LOS FM carrier 51:55.
19:00	DFE	Orroral DFE.
19:01	Opsr	Orroral.
19:02	DFE	Could you give me the time of the mod off on the FM dump?
19:06	REC	Roger 09:51:37.
19:09	DFE	Thankyou.
19:15	REC	LOS DFI 52:28.
19:21	REC	LOS OD 52:41.
19:32	DFE	Copy that.

END OF ORRORAL PASS RECORDING.

Tape Transcript Orbit 16 at approx. 22:15 local time 14 April 1981

Tape Time (Min/sec)	Callsign	Transcript
		Communications on external circuit
21:02	GO	Quito Goddard ops site coord.
21:05	Q	Quito
21:09	GO	OK Quito, I'm going to release you at this time, like to thank you for your support.
21:16	Q	Roger

Communications on air/ground circuits

21:16	HOUSTON COMTECH	OK you're load & clear, stand by.
21:20	Y	Roger.
21:25	HOUSTON COMTECH	(With tones) This is Houston comtech test one, two, three, two, one, end of test.
21:35	Y	Houston comtech Yarragadee comtech we have 100% keying modulation go.
21:39	HOUSTON COMTECH	Roger Yarragadee.
21:40	ORRORAL COMTECH	Orroral comtech we have 100% keying air/ground one air/ground two.
21:45	HOUSTON COMTECH	Roger Orroral, thank you. Configure for your pass.

Yarragadee talking on external circuit.

22:26	Y	Yarragadee is configured for real-time support,

22:28	HOUSTON COMTECH	Roger.

Communications on external circuit

22:40	RTC	Orroral RTC.
22:42	Opsr	We are configured for command interface.
22:36	RTC	RTC copies.

Yarragadee talking on external circuit.

22:56	Y	Yarragadee is go for voice.

SPACE SHUTTLE is through Yarragadee.

23:18	CAP COM	Good morning Columbia this is the Crimson Team through Yarragadee, we'll be with you for 8 minutes. How do you read? Over.
23:53	CAP COM	Good morning Columbia this is the Crimson Team through Yarragadee, how do you read over.
24:00	SPACE SHUTTLE	Good morning Crimson Team we read you loud and clear. How you doing Joe
24:03	CAP COM	OK good morning you are very, very weak. We've got nothing special for you. Except to say we are happy with the PCS config as it is right now.
24:17	SPACE SHUTTLE	Good morning Joe how you guys doing it's about time you all came to work.
24:22	CAP COM	We've just been watching and enjoying. We are proud of the Silver Team though. They did a grand job and so did you. We are thinking of having them bronzed, in fact.
24:46	SPACE SHUTTLE	(Garble)
24:49	CAP COM	OK, John and Crip, you are very weak, and we may have some comm problems if we don't get much to you this pass, we'll be back very shortly through Orroral Valley.

Yarragadee talking on external circuit.

25:08	HOUSTON COMTECH	Yarragadee comtech Houston comtech.
25:11	Y	Houston comtech Yarragadee comtech.
25:14	HOUSTON COMTECH	Are you having transmission problems?
25:15	Y	Negative. No I have no indication of problems, it is just weak signal.
22:24	Y	Would you like me to remove the squelch?
25:29	HOUSTON COMTECH	Negative Yarragadee, stand by.
25:32	Y	Roger.

SPACE SHUTTLE is through Yarragadee.

25:14	CAP COM	Columbia it is not comm problems, you are just weak in your transmissions to us.

25:48	SPACE SHUTTLE	(faint) it is ... looking at this...pressure . called the 02. Oh, you were happy with it. I missed that part.
26:04	CAP COM	OK. Crip. We are happy with the current PCS configuration. We are looking at that reg pressure and we will keep you advised on that.
26:15	SPACE SHUTTLE	Ok. You sure are easy to please.
26:21	CAP COM	Well, we may not be when we see some data here in a few minutes. But we are keeping a careful eye on it. There is nothing that can break as we watch it so we are not particularly worried.
26:35	SPACE SHUTTLE	It seems to have levelled off at around 215 or so.
26:40	CAP COM	OK. We copy that. Thank you.
28:12	SPACE SHUTTLE	I finally got around to my first cup of coffee. Sure tastes good
28:22	CAP COM	Roger that.
28:25	SPACE SHUTTLE	Not really a cup though.
28:45	SPACE SHUTTLE	... said something else.
28:49	CAP COM	Crip you are dropping out here.

Orroral & Yarragadee talking on external circuit.

29:20	DFE	Orroral DFE.
29:21	Opsr	Orroral.
29:22	DFE	Your lines are enabled would you confirm you are green for support?
29:26	Opsr	That's affirmative, do we have a clearance for destination code 150?
29:30	DFE	That's affirmative on that Orroral.
29:46	Opsr	Orroral destination code 150.
29:49	DFE	Copy that, we do see the 150, thank you.
30:36	Y	Yarragadee has one minute to LOS.

SPACE SHUTTLE is through Yarragadee.

| 30:46 | CAP COM | Columbia, Houston. We are 30 secs. from LOS. We will be gone for a minute and a half and be back with you at 23 + 22. |

Yarragadee talking on external circuit.

31:09	Y	Houston comtech Yarragadee comtech we copied no downlink.
31:13	HOUSTON COMTECH	Roger.
31:37	Y	Yarragadee has LOS.
31:41	HOUSTON COMTECH	Roger Yarragadee.

Communications on external circuit

33:06	REC	Orroral has acquisition OD 22:17.
33:10	DFE	DFE copies.
33:12	REC	Orroral has acquisition DFI 22:23.
33:20	USB	Orroral is go for command.
33:26	USB	Ranging initiated Orroral.
33:42	REC	Orroral has acquisition FM carrier 22:51.
33:46	DFE	DFE copies.
33:49	Opsr	Orroral range acq is complete, range delta is 0.9 microseconds.
33:57	REC	FM dump 22 er 23:07.
34:02	DH	PCM lock on dump FM, 960 reverse.

SPACE SHUTTLE is through Orroral

33:35	CAP COM	Hello, Columbia. This is Houston back with you through Orroral Valley. We will be with you for 3.5 min and can report that the IMU cal has been completed.
33:46	SPACE SHUTTLE	OK. Fine and dandy. You got any other traffic for us?
33:50	CAP COM	Not much, Crip. You are loud and clear on this pass. Curious to know if you have message 11 aboard. It is a pretty major change to timeline and prepare to answer questions, when and if.
34:05	SPACE SHUTTLE	OK. I'll tell you what. John is down on the mid deck now. We will check that out for us. Meanwhile I got a little Slim Dusty and Waltzing Matilda for our friends down under here.
34:16	CAP COM	Let her rip.
34:22	SPACE SHUTTLE	**(Music - 'Waltzing Matilda' sung by Slim Dusty– 37 seconds.)**
34:59	SPACE SHUTTLE	Too bad it is always dark when we are going over. We can't get a good view of it.
36:35	CAP COM	Aw, but they got it, a good sound of it. I think the S-band will never be the same again.
35:12	SPACE SHUTTLE	Probably not. Probably not.

Communications on external circuit

36:01	ORRORAL COMTECH	Houston comtech Orroral comtech site coord.
36:04	HOUSTON COMTECH	Houston comtech.
36:12	ORRORAL COMTECH	Roger, we copied a short burst of 450 HZ on air/ground 2. Its gone again now.
36:12	HOUSTON COMTECH	Roger.
36:23	HOUSTON COMTECH	Orroral comtech Houston comtech site coord

36:25	ORRORAL COMTECH	Orroral comtech.
36:27	HOUSTON COMTECH	You did copy Waltzing Matilda did you not on the air to ground?
36:31	ORRORAL COMTECH	That's Affirmative, we almost recognised it.

(Explanatory Note: The complex sounds of the music were very badly distorted by the severe limitations of the air to ground voice link between the spacecraft and the ground. This link was specifically designed for voice communications only.)

36:40	HOUSTON COMTECH	Ha, Ha, roger that.
36:53	REC	FM dump off 25:58.
36:57	DFE	Thankyou.
36:58	Opsr	Approximately 1 minute to horizon LOS.
37:14	REC	LOS FM 26:17.

SPACE SHUTTLE is through Orroral

37:17	CAP COM	Columbia, we are about 30 seconds from LOS. We'll be back with you at 23 plus 54.
37:26	SPACE SHUTTLE	Roger, 23:54.
37:29	CAP COM	And we enjoyed the music, Bob, Thank you.
37:32	SPACE SHUTTLE	Oh, we enjoyed it. We just wanted to share some with you.

Communications on external circuit

37:51	REC	LOS DFI 26:55.
37:56	DFE	DFE copies.
38:02	REC	LOS OD 27:06.
38:07	DFE	Copy that Orroral.
38:24	Opsr	GC Orroral.
38:27	GC	Orroral GC.
38:29	Opsr	We'd like to convey our thanks to the Columbia crew for the little bit of hometown music.
38:35	GC	Well, OK We'll pass that on.

END OF ORRORAL PASS RECORDING

(GC is the Houston operations controller.)

Tape Transcript Orbit 17 at approx. 23:53 local time 14 April 1981

Tape Time (Min/sec)	Callsign	Transcript
38:51	CAP COM	The vectors on board now.
38:52	SPACE SHUTTLE	OK that was fast.

Communications on external circuit

Tape Time (Min/sec)	Callsign	Transcript
38:59	Opsr	DFE Orroral
39:00	DFE	Orroral DFE
39:03	Opsr	Did you copy that the FM came on with modulation?
39:07	DFE	Yes, I did thank you and you are seeing the 5 to 1 forward.
39:11	Opsr	That's affirmed.
39:12	DFE	Roger. We had a drop just prior to that on the OD. Do you have any indication of what caused the drop it would indicate we missed? We just had a drop in the OD.
39:21	Opsr	Stand by, we'll check.
39:25	Opsr	Telemetry reports they didn't see a drop.
39:41	Opsr	DFE Orroral.
39:44	DFE	Orroral DFE.
39:45	Opsr	We just did copy a momentary OD drop.
39:49	DFE	OK. OK we just dropped again.
39:54	Opsr	We copied that one as well.
39:56	DFE	Roger.

END OF ORRORAL PASS RECORDING

Tape Transcript Orbit 18 at approx. 01:28 local time 15 April 1981

Tape Time (Min/sec)	Callsign	Transcript

Communications on air/ground circuit

Tape Time (Min/sec)	Callsign	Transcript
40:18	ORRORAL COMTECH	Orroral comtech air/ground 2 load & clear.
40:21	HOUSTON COMTECH	Roger you're load & clear, Yarragadee comtech Houston comtech air to ground 2.
40:28	Y	Houston comtech Yarragadee comtech

| 40:33 | HOUSTON COMTECH | Standby for a keying check. (with tones) |
| 40:39 | HOUSTON COMTECH | Houston comtech test one, two, three, two, one, end of test. |

Communications on external circuit

40:39	DFE	Ascension DFE.
40:42	ASC	Ascension, go ahead.
40:43	DFE	I understand you've already been checked out by RTC, so I will turn your lines off at this time.
40:40	ASC	Roger.

Communications on air/ground circuits

| 40:51 | ORRORAL COMTECH | Houston comtech Orroral comtech Go ahead Yarragadee. |
| 40:57 | HOUSTON COMTECH | Orroral meet me on site coord. |

Communications on external circuit

41:02	ORRORAL COMTECH	Houston comtech Orroral comtech site coord.
41:05	HOUSTON COMTECH	Ok Orroral you see your comm configuration Lima?
41:08	ORRORAL COMTECH	That's affirmative.
41:11	HOUSTON COMTECH	Stand by and we'll put teleprinter on and let you check your levels.
41:17	ORRORAL COMTECH	Roger.
41:42	DFE	Hawaii DFE voice check,
41:45	HAW	I read you load & clear.
41:46	DFE	I read you fives also Hawaii, are you ready to er go for your sequence 25?
41:53	DFE	OK Hawaii, (garble).
42:13	RTC	Hawaii RTC.
42:17	HAW	Hawaii.
42:18	RTC	This is RTC, can you pick up on sequence 35?
42:25	HAW	Roger stand by.
42:27	RTC	RTC copies.

Communications on air/ground circuits

41:53	HOUSTON COMTECH	Orroral comtech Houston
41:59	ORRORAL COMTECH	Houston comtech Orroral comtech
42:01	HOUSTON COMTECH	Go ahead.

42:04	ORRORAL COMTECH	Roger, 2056 at neg 13. Just standby we'll set levels at the DMS.
42:38	ORRORAL COMTECH	Houston comtech Orroral comtech
42:41	HOUSTON COMTECH	Go ahead
42:42	ORRORAL COMTECH	Roger, we have a neg 22.5 at the DMS and a verification level of neg 20.
42:49	HOUSTON COMTECH	Roger Orroral, configure comm configuration Lima.
42:54	ORRORAL COMTECH	Roger.
42:56	HOUSTON COMTECH	Thankyou.

Communications on external circuit

42:57	HAW	RTC Hawaii. I confirm 1024 reverse on OD high bit rate. Also configured for sequence 35.
43:35	Opsr	RTC Orroral.
43:11.	RTC	Orroral RTC
43:12	Opsr	We're configured for H-10 interface.
43:14	RTC	Roger stand by we'll be with you in just a second Orroral.
43:18	RTC	Hawaii RTC.
43:19	HAW	Hawaii.
43:20	RTC	You are go on your sequence 39, safe your SCVM please.
43:25	HAW	Roger.
43:27	RTC	Orroral RTC.
43:29	Opsr	Orroral.
43:30	RTC	Stand by.
43:33	DFE	Hawaii DFE, I er confirm your PM & FM are both fives. Would you give me the DFI at this time please?
43:44	HAW	Roger.
43:46	Y	Houston comtech Yarragadee comtech.
43:40	HOUSTON COMTECH	Houston comtech.
43:52	Y	Yarragadee configured for real-time.
43:54	HOUSTON COMTECH	Roger.
43:56	RTC	Orroral RTC.
43:58	Opsr	Orroral.
43:50	RTC	You are go on your H-10, remove your modulation & configure for support please.
44:04	Opsr	Wilco

44:12	ASC	Houston Track Ascension
44:16	DFE	Hawaii DFE, I see good DQM on your DFI. That completes the interface you are green. You can go ahead and take your data generator off line and normal up.
44:25	HAW	Roger.
44:26	DFE	Thank you.
44:27	HAW	You're welcome;
44:32	DFE	Ah, I will turn your lines off Hawaii, you might like to DLSM at this time.
44:38	HAW	Roger.
45:08	ASC	Track Ascension site coord.
45:36	Y	Houston comtech Yarragadee go for voice.
45:39	HOUSTON COMTECH	Roger.

SPACE SHUTTLE is through Yarragadee.

45:46	CAP COM	Hello, Columbia, this is Houston back with you through Yarragadee for six minutes. How do you read?
45:53	SPACE SHUTTLE	Roger, Fine, Joe. That's just what one of my little computers told me that I had you for six minutes.
46:00	CAP COM	Okay. They never lie.
46:07	SPACE SHUTTLE	Well, we're sitting here in OPS 3 all squared away just a little bit early. Took our burn out of the road. We got the radiators all stowed.
46:27	CAP COM	Okay, Crip. We copy that and if you've got time on your hands, and a pad handy, I can read up to you the RCS test sequence number 3.
46:47	SPACE SHUTTLE	Stand by one, Joe.
46:49	CAP COM	ROG. No hurry.
46:52	SPACE SHUTTLE	Well, I've just been getting a pen and getting organised. Okay, I just got it. Both John and I standing by to copy.
46:57	CAP COM	Okay, RCS test sequence number three. The burn attitude solution numbers are 342, 136, and 323. Target 144 by 144. Delta V 2.6 and tgo is 05. It's a multi-burn axis—make that a multi-axis burn with a footnote do the burn in your RCS 2 post burn attitude. The RCS 2 post-burn attitude. The weight 207480 Tig 001/034200.0. Delta V's are plus 0.7, all balls and minus 2.5. The post-burn att 311.2 182.7 and 013.0. The end attitude time is 4 hours and 44 minutes, over.
48:36	SPACE SHUTTLE	OK Joe coming back at you attitude three fourths solution attitude 342, 136, 322, 144 by 144 delta V total is 2.6 that is 0.3 seconds multi-axis burn with thr burn in the post RCS 2 postburn attitude weight is 207480 tig time is one day 3 hours 42 minutes 0 seconds delta V are +0.3 all balls and -2.5 the in time, correction, the postburn att is 311.2, 182.7, 013.0, in time is 444, over.

49:34	CAP COM	Ok Crip three corrections the burn attitude solution in the yaw is 323, 323, the Tgo is 5 seconds
49:53	SPACE SHUTTLE	323
49:53	CAP COM	Roger that. Tgo is 5 seconds and the last correction the delta VX is +0.7 over

Communications on external circuit

50:11	Y	Houston comtech Yarragadee comtech.
50:13	HOUSTON COMTECH	Go ahead.
50:14	Y	We had no uplink on that last message.
50:17	HOUSTON COMTECH	No uplink?
50:18	Y	That's er, no key.
50:21	HOUSTON COMTECH	Roger.
50:23	Opsr	DFE Orroral.
50:26	DFE	Orroral DFE.
50:27	Opsr	We have 3 mins to AOS your lines are not on.
50:30	DFE	Copy that Orroral, going on at this time.

SPACE SHUTTLE is through Yarragadee.

| 50:32 | CAP COM | Columbia this is Houston over. |
| 50:35 | SPACE SHUTTLE | Ok Joe a couple of those I already (garbled) ran back down on the yaw for the burnout is 323 the delta VX was 0.3 what else do you have. |

Communications on external circuit

50:36	HOUSTON COMTECH	Yarragadee Houston, is it clear now?
50:39	Y	That's affirmative
50:43	Opsr	DFE Orroral, are we clear for a destination code of 150?
50:47	DFE	That is affirmative.

SPACE SHUTTLE is through Yarragadee.

50:51	CAP COM	+0.7, 0.7
50:52	CAP COM	Roger that Crip and the Tgo is 5 seconds. Ok Crip the Tgo is 5 seconds over.
50:54	SPACE SHUTTLE	(garbled)
51:54	CAP COM	OK Crip the Tgo is 5 seconds, over,
51:08	SPACE SHUTTLE	(garbled)
50:55	Opsr	You have 150.

Communications on external circuit

51:11	HOUSTON COMTECH	Yarragadee Houston comtech
51:13	Y	Yarragadee.
51:16	HOUSTON COMTECH	Did that go up?
51:17	Y	That's affirmative and we are 1 minute from horizon.
51:20	HOUSTON COMTECH	Roger

SPACE SHUTTLE is through Yarragadee.

51:33	CAP COM	Columbia this is Houston we are having trouble copying you, no com problem but we need a little stronger signal which we will get Orroral Valley in a couple minutes and we'll finish the corrections there. Over.
51:50	SPACE SHUTTLE	Garbled.

Communications on external circuit

52:26	Y	Yarragadee LOS.
52:20	HOUSTON COMTECH	Roger.
53:20	REC	Orroral has AOS OD.
53:26	DFE	DFE copies.
53:27	REC	We have AOS DFI.
53:30	USB	Orroral is go for command.
53:36	RTC	RTC copies.
53:41	USB	Orroral has initiated ranging.
53:52	REC	Orroral has FM on.

SPACE SHUTTLE is through Orroral.

53:55	CAP COM	Columbia this is Houston over.
53:57	SPACE SHUTTLE	Hello Joe how do you read now.
54:00	CAP COM	Ok Crip you are loud and clear we're going to send some tin boost to you on water quantities this pad.

Communications on external circuit

54:00	Opsr	Range acquisition is complete. Delta range is 2.5 microseconds.

SPACE SHUTTLE is through Orroral.

54:06	SPACE SHUTTLE	Ok fine I'll read you back those corrections you gave me a while ago. The yaw attitude was 323, the tgo was 05 and the delta VX was +0.7.

Communications on external circuit

| 54:16 | DH | Orroral has PCM lock on 960 kilobit reverse data. |

SPACE SHUTTLE is through Orroral.

| 54:21 | CAP COM | Ok Bob that's correct and the end attitude time we think is right but its 4 hours and 44 minutes over. |
| 54:31 | SPACE SHUTTLE | Roger its 4 hours and 44 minutes. |

Communications on external circuit

| 54:34 | DFE | DFE copies your last Orroral. |

SPACE SHUTTLE is through Orroral.

| 54:44 | CAP COM | Ok Columbia we are showing you about 7 degrees out in yaw and it's an RCS burn coming up and we are showing OMS selected at the moment. |
| 54:58 | SPACE SHUTTLE | OK we yes your right your right. Thank you. |

Communications on external circuit

54:58	DH	Orroral had a drop in the 960 kilobit, we still have reverse data at this time.
55:04	DFE	DFE copies.
55:16	CAP COM	And that yaw fixed itself on the reload for the RCS.
55:21	SPACE SHUTTLE	Roger that.
55:31	CAP COM	Ok Columbia it looks right on the money to us. We're with you for 2 more minutes here.

Communications on external circuit

55:43	OPSR	Track Orroral.
55:48	T	This is Houston track
55:50	Opsr	OK We dropped range, we have re-acquired.

SPACE SHUTTLE is through Orroral.

| 55:44 | CAP COM | John and Bob, the teleprinter message you might hear rattling on its way to you is DFI recorder troubleshoot procedure. |
| 56:05 | SPACE SHUTTLE | OK Thank you |

Communications on external circuit

56:01	Opsr	range acq is complete, range delta is 3.6 microseconds.
56:18	T	Orroral Track on site coord.
56:21	Opsr	Orroral.

| 56:22 | T | Ok we see your good data now after you re-acquired. |

SPACE SHUTTLE is through Orroral.

| 56:24 | CAP COM | Roger. And Columbia, on that message. After you have had a chance to read it, we indicate what talk-back should read several times through it. It is not clear that it will read that, and in any case, proceed on with that procedure. Go right ahead with it. The talk-back positions are just hopes on our part. |
| 56:50 | SPACE SHUTTLE | OK. Joe is there something you want us to proceed with in AOS or when we get it and understand it and we got a go to do it? |

Communications on external circuit

57:15	Opsr	RTC Orroral you have one minute to horizon.
57:23	RTC	RTC copies.
57:26	REC	Orroral has LOS dump data.
57:30	DFE	DFE copies Orroral.

SPACE SHUTTLE is through Orroral.

57:31	CAP COM	OK. Columbia RCS 2 you want to be in roll 048.6 169 and 304.2
57:45	SPACE SHUTTLE	Stand by one. You are implying we are not in the correct attitude.
57:49	CAP COM	That's affirm. It is a multi-axis burn. You want to be in that post burn attitude which is 48.6, 169 and 304.2.
58:02	SPACE SHUTTLE	OK Joe. I guess we are confused. I guess we did not understand that you wanted us to go to the post burn attitude. You want us at the post burn attitude right now?
58:13	CAP COM	Multi-axis burn always uses post burn attitude.
58:20	CAP COM	Columbia we are in the blind. If you can't do it that way, do it in plus X.

Communications on external circuit

58:32	REC	Orroral has LOS all downlinks.
58:34	DFE	DFE copies that.
59:10	DFE	Orroral DFE would you configure for post-pass.
59:12	Opsr	Wilco.
60:04	Opsr	Goddard Ops Orroral.
60:07	GO	Go ahead Orroral.
60:08	Opsr	This is the last pass of this series, we'd like a release on the RF equipment when you can arrange it please.
60:17	GO	OK. You'll still have playback, we'll see what we can do for you.

-

60:22	Opsr	OK. Thankyou
60:25	HAW	Track Hawaii we are standing by for sequence 42.
60:49	HAW	Track Hawaii.
60:51	Track	This is Houston track, go ahead.
60:53	HAW	Be advised we are standing by at sequence 42.
60:56	Track	Alright, go ahead and flow your data.
61:07	HAW	Low speed data on line.
61:17	Opsr	DFE Orroral.
61:21	DFE	Orroral DFE.
61:22	Opsr	We're configured post-pass.
61:24	DFE	Give me a type 1 command history on your mark. Your lines are coming on at this time.
61:31	SCE	Mark, command history.
61:33	Opsr	Command history is complete.
61:35	Track	Hawaii this is Houston track, we see your data.
61:42	HAW	Roger,
61:47	DFE	Orroral DFE that completes my requirements, would you start queuing up for the dump playback and I'll put you on the playback coord loop.
61:53	Opsr	Roger.
61:55	Track	Hawaii Houston track, we have seen enough of your low speed tracking data. You can turn it off and prepare for pass.
62:01	HAW	Roger.
62:06	HAW	Low speed is terminated.
62:08	GO	Orroral Goddard Ops on site coord.
62:11	Opsr	Orroral.
62:13	GO	I give you a 14:39 on release of your RF equipment only.
62:16	Opsr	Thankyou

END OF RECORDING.

www.ingramcontent.com/pod-product-compliance
Lightning Source LLC
Chambersburg PA
CBHW040141200326

41458CB00025B/6342